W9-BZO-130

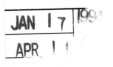

TOXIC NATION

TOXIC NATION

The Fight to Save Our Communities from Chemical Contamination

FRED SETTERBERG
LONNY SHAVELSON

Photographs by
Lonny Shavelson

John Wiley & Sons, Inc.
New York • Chichester • Brisbane • Toronto • Singapore

Copyright © 1993 by Fred Setterberg and Lonny Shavelson
Published by John Wiley & Sons, Inc.
Photographs © Lonny Shavelson

Library of Congress Cataloging-in-Publication Data:

Setterberg, Fred.
    Toxic nation : the fight to save our communities from chemical contamination
  Fred Setterberg, Lonny Shavelson.
        p.   cm.
    Includes index.
    ISBN 0-471-57545-3 (acid-free paper)
    1. Environmental health—North America. 2. Pollution—Social aspects—North America.
  I. Shavelson, Lonny. II. Title.
    RA566.5.N7S48   1993
    363.73′8′0973—dc20   93-8010

Printed in the United States of America

10 9 8 7 6 5 4 3 2 1

To Sandy Close

CONTENTS

ACKNOWLEDGMENTS

Many people assisted in innumerable ways in the writing of this book. In particular, we owe much thanks and appreciation to: Mike A'Dair; David Beers; Deborah Bickel; Elinor Blake; Susan Brenneman; Tom Brom; Zelda Bronstein; Mary-Helen Burnison; A. D. Coleman; Gena Corea; Gary Delgado; Jeannie Echenique; Renee Emunah; the emergency department staffs of Kaiser Hospital, Walnut Creek, and Alta Bates Hospital, Berkeley, California; Sue Ferris; Ping and Carol Ferry; Loralie Froman; Fred Goff, Andy Kivel, and the staff of the Data Center in Oakland, California; John Heilborn; the Fund for Investigative Journalism, Washington, D.C.; Les Hodge; Impact Visuals in New York; Carol Jenkins; Jane Kay; Marshall Krantz; our agents Michael Larsen and Elizabeth Pomada; Wendy Ellen Ledger; Maureen Murdock; Emory Menefee; Peggy Northrop; April Rapier; Amy Shapiro; Peter Torsiello and the staff of the Coffee Mill; Ann Van Steenberg; Susan West; Lyle York.

Sincere appreciation to Sandy Close, editor of Pacific News Service, who encouraged this project from its beginnings, and to Kate Bradford, our editor at John Wiley & Sons, Inc., whose patience and keen judgment helped bring about its completion.

Special thanks to Julian Price.

PROLOGUE

"People at the grassroots level feel it in their hearts. They see it. They hear the evidence. And they're getting a little frustrated that many world political leaders just don't see it."

—Vice President Al Gore

One month before taking office, Vice President-elect Al Gore scored front-page headlines by offering a dramatic clue to the incoming administration's environmental policies.

"Serious questions concerning the safety of an East Liverpool, Ohio, hazardous waste incinerator must be answered before the plant may begin operation," Gore declared. Citing his concerns about "the safety and health of local residents," America's first environmental vice president promised there would be a full investigation by the General Accounting Office before the new administration would allow the test-burn and start-up of the toxic waste incinerator, scheduled for the following month.

Events in East Liverpool seemed to indicate a major shift in national environmental policies. But most reports completely missed the larger story taking place in this conservative small town.

For 11 long years before Gore had arrived on the scene, a small army of parents, teachers, physicians, nurses, steelworkers, factory hands, retirees, and other townspeople had successfully blocked the start-up of East Liverpool's hazardous waste incinerator. While the press and public understandably focused on the new administration's environmental policies, the 11-year battle that had already stopped the incinerator and led to the vice president's dramatic statement remained untold.

Enraged by the prospect of lead, mercury, dioxins, and other toxic chemicals pouring into the air 1,100 feet from East Liverpool's elementary school, citizens had climbed fences to stop the construction of the hazardous waste burner, blocked the entrance gates when toxic wastes arrived at the plant, fasted for 47 days in protest, and finally boarded a bus for Washington, D.C., to sit in at the office of then-EPA chief William Reilly—demanding an audience, until the police hauled them off to jail.

For 11 years, residents of the Ohio, Pennsylvania, and West Virginia corner of the Ohio River had formed study groups to learn about the dangers of dioxins and lead. They had researched the scientific basis of the incomplete burning process inherent to "state of the art" hazardous waste incinerators. They visited other communities across the country where citizens claimed high rates of illness from toxic waste burners, and they used the Freedom of Information Act to secure access to confidential licensing files belonging to the EPA and the incinerator company. In the end, they went to court (paying for their legal costs with income from spaghetti dinners and garage sales) and obtained restraining orders against the incinerator company.

At one point, when their legal battle seemed finally to run aground, the people of East Liverpool simply hunkered down in front of the plant and said: *No.* On Monday, a group of grandparents gathered, linking arms, and blocked the incinerator's gateway. On Tuesday, it was the parents' turn. Wednesday, health professionals. Thursday, small business owners. Friday, steelworkers. On weekends, the groups joined forces, disrupting work at the plant until a new legal maneuver produced another restraining order, and once again stopped the plant's test-burn.

A local newspaper poll reported that an astounding 72 percent of residents surveyed approved of further civil disobedience to stop the storage and burning of hazardous chemical wastes at the facility. Many people in town recognized that they were desperately plugging up a leaky dike with their fingers—but for 11 years, their efforts made the dike hold.

Gore's highly visible intercession was widely regarded as the first sign of resurgent environmental activism at government's highest level. The Bush administration's environmental policies of evasion and equivocation were over. Even Reagan's strident opposition to almost any kind of environmental regulation now seemed chiefly a dim, if dismaying memory. The citizens of East Liverpool had sustained their efforts through the worst of times. Now that Gore and the new administration had come to power, couldn't they take a rest?

In fact, the opposite proved true.

As of February 1993, when the incinerator was scheduled to open, the new administration had not yet taken action. In desperation, local residents filed one more lawsuit. They obtained another restraining order, citing unresolved safety and licensing issues. And they reminded Vice President Gore of his promise to halt the toxic waste burn, pending the General Accounting Office's investigation.

"We were successful in delaying the test-burn," East Liverpool resident and nurse Terri Swearingen wrote to the vice president. "We carried the ball for you." Swearingen, who lived two miles downwind from the incinerator, had been arrested five times for blocking the operation's start-up. She had no intention of giving up now.

"Since the new Democratic administration came out with its state-ment that they'd stop this toxic waste burner," she explained, "I've been busier than ever. You have to hold the politicians' feet to the fire—otherwise, they may conveniently forget you. I really think Gore wouldn't even have known about this place if law-abiding, God-fearing people weren't spending time in jail—making a sacrifice now, so that our children aren't sacrificed later."

In thousands of towns across the country, a powerful grassroots anti-toxics rebellion like the one staged in East Liverpool had sprung up, hidden from the public eye. Faced with a new administration that might listen to their concerns about their families' health and safety, the people who composed this burgeoning but largely invisible small-town anti-toxics movement were redoubling their efforts, demanding attention and action.

No matter who controlled the White House, they had learned to depend, first and foremost, on themselves. The 1990s, they hoped, would be their decade.

Fred Setterberg and Lonny Shavelson
March 1993

TOXIC NATION

Into the Heart of Toxic America

Nova, Ohio.

Mr. Keim seemed the model of Amish patience and propriety.

The full-bearded patriarch of 11 children and 82 grandchildren sat upon a hard-plank, cushionless stool in the common room of his two-story wood-frame house in north-central rural Ohio. He smiled queerly—his front teeth were missing—and waved us closer to the fuming wood stove. He wore rough home-sewn laborer's garments; a plain grey stitched chemise blue denim trousers, handmade woolen socks, and heavy leather work boots. Mr. Keim's daily life, like that of the other 140 Amish farming families who had settled this region four decades earlier, revolved around the parochial concerns of commu-

nity, faith, and family bound together by a seventeenth-century doctrinal insistence "to be in this world, but not of it." Withdrawn from
modern society, the Amish thrived in a manner largely unchanged
over the past three hundred years. But now, Mr. Keim felt certain that
the Amish way of life faced an unprecedented disruption.

International Technologies, the huge waste disposal company,
wanted to construct a toxic waste incinerator in the countryside bordering the Amish farms. The fact that Mr. Keim's minister, the most
powerful individual in the community, had selected him to usher into
his home two reporters to explain the objections of the Amish indicated the gravity with which they regarded the matter.

"If you want to know how the Amish feel," urged Mr. Keim, his
snowy hair bundled at the ears, setting off steel grey-blue eyes that
seemed to glow as he told us his story, "*read this.*" We fitted ourselves
into a pair of straight-back walnut chairs, scooted closer to the fire,
and studied the inflammatory Greenpeace leaflet that had fluttered
from Mr. Keim's outstretched hand on to the table. A more unlikely
broadside could not have been found among the insular Amish.

"*We don't know what they were burning in their incinerator,*" read the
flyer, quoting the anxious complaint of Mary McCastle, a black woman
from Alsen, Louisiana, where another toxic waste incinerator had
been built, "*but we know that it was making us sick. We know that we
couldn't hardly have rest in our own homes. We couldn't have any more beautiful gardens. . . .*"

An immense cultural gulf separated black rural Louisiana from
Amish Ohio. But Mr. Keim wasn't interested in the differences. He
could see from the leaflet that the two communities had identical
worries. The Amish feared that the fumes, smoke, residue, ash, or
wind-borne particles of incinerated toxic wastes might flit across the
skies to settle upon their farms and contaminate their lives. In Louisiana, Mary McCastle asserted that it had already happened to her
people.

Was America really poisoning its own citizens?

Over the past three years, we had listened to hundreds of people
confide, insist, rail, and worry that the unfettered proliferation of
toxic wastes had devastated their lives. We had crisscrossed the country, traveling from California to Massachusetts, Pennsylvania to Mississippi, visiting more than two dozen states to talk with people in
cities, small towns, and rural hamlets who believed that their families
were imperiled by a pervasive menace that had been churned out into
the environment by hazardous waste dumps, dioxin-spewing industrial chimneys, toxic waste incinerators, pesticide-spraying airplanes,
home garden weed killers, apartment buildings saturated with for-

maldehyde, chemical food contaminants, leaking landfills, legal and illegal dumping, and industrial accidents. This confrontation with an invisible invader had reordered thousands of lives, turning ordinary citizens into a motley procession of victims, rebels, instant experts, and slow-boiling activists—heroes or hysterics, depending on your point of view. The popular vision of the post–World War II petrochemical miracle had blurred into a fractious portrait of resentment, betrayal, and rage.

But were the fears justified? Did the nation's mounting concern over toxic wastes have a rational basis? Or rather, did the toxics crisis constitute a vast national exaggeration that diverted time, attention, political will, and billions of dollars from far more urgent problems? Moreover, how had toxics affected the way Americans thought about their country, their future, their lives? What were the cultural implications of inhabiting a "poisoned" world?

These questions led us for three years across an American landscape that we never could have previously imagined. We began our journey in McFarland, California, a small farming community in the Central Valley where the cancer rate among children had soared 400 percent above the national average. Nobody knew why. Many local parents felt certain that the children's cancers—one child stricken on almost every block in town—came from exposure to the pesticides sprayed upon nearby fields, or some other unknown environmental contamination. Yet the state health department officials could find no evidence supporting these theories. In fact, health officials initially believed the McFarland cancer cluster might simply be a statistical fluke, an aggregation of bad luck.

And perhaps they were right. In time, McFarland might be classified as a medical mystery whose complex linkage of misfortune indicated no significance for the larger world. But if the official story turned out to be wrong—or incomplete, as growing numbers of people had come to believe—the entire nation would have to deal with the consequences.

As we began to look beyond McFarland to other communities throughout the nation with similar problems, another possibility ineluctably took shape. Rather than a mere fluke, the sad fate of McFarland's children might possibly signal a larger, far more menacing health hazard. To better understand what was happening throughout the country, we focused our attention on the small towns and cities where growing numbers of people were increasingly convinced a toxic world had made them ill.

We wanted to learn why so many people believed that their lives and communities were now poisoned, what living in a "toxic environ-

ment'' actually meant to them—and finally, most importantly, did our
nation now face a growing health threat triggered by massive chemical
contamination?

One thing was certain: America's toxics overload had spread far
beyond the narrow boundaries presumed by most citizens.

The U.S. government estimates that over sixteen thousand active
landfills have been sopped with industrial and agricultural hazardous
wastes. Most are located near small towns and farming communities—
and the contents of all of them, according to the Environmental Protection Agency (EPA), will eventually breach their linings and penetrate the soil, as many already have done. Underground chemical and
petroleum storage tanks scattered throughout cities, suburbs, and
rural America number between three and five million; 30 percent
already leak. Pesticides have contaminated water supplies in 23 states,
leaching into aquifers and washing into streams and rivers where they
end up in the water we drink and the fish we eat. According to industry's own reports, 22 billion pounds of toxic chemicals are spewed
into the air, water, and soil each year—about 85 pounds of toxic waste
for every American. The Congressional Office of Technology Assessment estimates the real figure to be vastly higher.

The full extent of the nation's toxic contamination may not be
understood for decades to come. EPA officials have identified 32,000
sites throughout the country that require monitoring to determine
whether their threat to living creatures warrants upgrading to Superfund status, the program charged with cleaning up the most severe
hazardous waste hot spots. And yet, according to the General Accounting Office, the EPA "does not know if it has identified 90 percent of
the potentially hazardous waste sites or only 10 percent." There may
be as many as three hundred thousand toxic waste sites spread
throughout the nation—one for every nine hundred Americans. Millions of people live near, next to, or even on top of these sites, but
the present and future effects of spending years within proximity to
their contents remain uncertain.

Given the limits of science, nobody can accurately predict which
sites, if any, will increase cancer rates, depress immune systems, spawn
chemically induced learning disorders among children—or simply
transform the life of the community by the dread and uncertainty the
sites produce in vast and immeasureable quantities. Given the limits
of government oversight and political conviction, few people in power
have been willing to squarely confront the unbridled culture of consumption that makes inevitable the incessant overflow of dangerous
wastes.

That is why we now found ourselves in Amish Ohio, seated around the crackling wood stove in the common room of Mr. Keim's house.

"When Adam and Eve were driven out of the garden," explained Mr. Keim genially, continuing his lecture without rancor or agitation, "they were supposed to farm the earth and live from it. They had, as we do, some problems to contend with. But if they polluted the earth to the extent where they couldn't farm, well, that would be. . . ."

Mr. Keim trailed off into an ellipsis of the obvious, his cupped hands opening upon his knee as though releasing a small bird. It would be, he was indicating, an abomination. The chemical contamination of the region's rich farmland would mean economic catastrophe and the devastation of a way of life.

The Amish were not alone in their fears.

Other farmers working the land nearby, who were long accustomed to using modern chemical fertilizers and pesticides, shared the Amish skepticism about the incinerator's safety. They put aside any misgivings they may have had about Amish eccentricities in dress and custom, and requested that their neighbors join the larger community in urging the government to block International Technologies' bid to build its incinerator.

Of course, the farmers knew that public action by the unworldy Amish seemed unlikely. But when Ohio senator Howard Metzenbaum arrived to weigh his constituents' arguments against the incinerator, the local farmers chauffeured the senator into the countryside to inspect its proposed location—and the entire party encountered an astounding sight. As they drove along the two-lane country road, the senator's car passed rows of Amish men sitting on the hillsides in respectful silence upon the bedboards of their horse-drawn wagons. They didn't hold signs or banners, or even speak to the senator. But it was clear why they had assembled. The Amish presence along the road was read as an unmistakable sign that passions over the incinerator could not be ignored. If the reclusive sect had been incited to act, there would certainly be wider political consequences throughout the region.

And yet, once we had left Mr. Keim's plain and simple home and abandoned the warmth of his potbelly wood stove, it occurred to us that Amish involvement in the toxics controversy might cut two ways.

To some people, the gathering of Amish farmers dressed in their formal black suits and full beards must have seemed like a visitation from the pre-industrial past, a ghostly warning from a safer, saner world regarding the excesses of modernity and the dangers of rampant progress.

To other observers, the Amish presence would have appeared a masterpiece of ignorance. The archaic sect's stubborn unworldliness might stand as the most extreme example of the modern Luddites whose intransigence only muddled the best efforts of science and government to solve the toxics dilemma. Did it really make sense to read wisdom into the actions of a community that also saw the use of transistor radios or electric fans as a breach of their heavenly covenant? Who could we believe when it came to answering the question of whether toxic contamination was undermining the nation's health? And if fear of toxics had penetrated the well-armored Amish community, how much deeper and wider had its impact spread, in various ways, throughout the rest of American culture?

"Do you really think you'll be able to win your fight?" one of us asked Mr. Keim, as we stood at his doorstep shaking hands goodbye.

"Sometimes with a lot of power and authority," he whispered, "the Lord will turn people around."

This book is the story of people whose lives have been turned around decisively by their belief that our nation is now being engulfed by its own poisonous excess. Our task has been to understand how their apocalyptic vision originated and then quickly spread—and what it now means for the rest of us.

But this narrative is not simply a chronicle of environmental neglect and potential health hazard. Rather, our exploration of toxic America inadvertently provided a view of our society as it struggles with the profound divisions that arise whenever we confront grave questions about the future. Over time, the psychological, social, and political implications of the nation's expanding roster of "contaminated communities" have become as important to us as the medical and scientific mysteries around which their identities are formed. This book isn't exclusively about poisoned people; ultimately, it deals with how our nation contends with its most fundamental problems.

"The Whole Neighborhood Was Stunned"
—McFarland, Fall 1987

McFarland, California. Tina Bravo and her son.

Tino Bravo could not see anything beyond the front porch of her home.

The dense tule fog had descended upon California's Central Valley, enshrouding the entire community of McFarland in a thick, slate-grey haze. During the long dark mornings and bright blinding winter afternoons, the cars speeding along U.S. Highway 99 sometimes failed to anticipate the hazards of the low-lying fog—and as a result, disaster ensued. Passenger cars mashed their brakes and then skidded into slow-moving pickup trucks; buses scrambled on top of the lumbering produce vans and U-Hauls.

Eventually some enormous interstate-bound eighteen-wheeler would scream up blindly from behind, flattening the entire procession into an accordian wreck of screeching metal. Road accidents involving 15, 25, 30 vehicles were not uncommon. Winter fatalities peaked around March, before the warmer temperatures offered a brief respite until the descent of the thinner, but equally lethal, summer fog. You couldn't see, the drivers always explained, you couldn't see where you were or where you were going.

U.S. Highway 99 bisects McFarland, dividing the town into two flaps of stucco tract-home subdivisions. Tina Bravo owned a three-bedroom home on the eastern flap, the newer side of town built in the late 1970s with the aid of federal subsidies for low-income residential developments. They were modest homes, the kind found in thousands of other blue-collar communities throughout the country. Like most of her neighbors, Bravo felt lucky to own something so solid, serviceable, and new.

On this particular Sunday morning, February 16, 1987, Tina Bravo stood at her front door, arms folded, her face taut and strained from squinting into the dense grey mist; she was attempting to puzzle out the identities of the figures passing in front of her house.

Bravo had lived in McFarland since 1980, and she knew the town well. McFarland is located about 160 miles northwest of Los Angeles in the San Joaquin Valley section of California's Central Valley—one of the world's richest farm belts. About 35 miles south of McFarland—still Kern County—lies the town of Weed Patch, the site of the fictional migrant camp that housed the Joad family in Steinbeck's *The Grapes of Wrath*. Delano, only five miles north, is the birthplace of Cesar Chavez's United Farm Workers union, which successfully organized field laborers in the 1960s.

McFarland's dozen or so blocks of tract homes are bordered on all sides by fields of cotton, almonds, kiwi fruit, and grapes. The majority of the six thousand two hundred people who live in McFarland are Mexican-American. Most of the families descend from farm laborers. Even today, almost all of the town's residents earn their livelihood from occupations financially tied to the valley's extraordinarily productive soil. The highway billboard on the outskirts of town proudly proclaims: "McFarland: The Heartbeat of Agriculture."

Although McFarland's residents have always shared the small town concern for privacy, Tina Bravo thought she understood her neighbors at their most basic level. She particularly knew what bothered, delighted, inspired, and frightened McFarland's parents. And on this Sunday morning in February, after the newspaper hit McFarland's porches and word had spread throughout the community, Tina Bravo knew that practically every parent in town was shaken by the same thought: Thank God it wasn't *my* kids.

McFarland's children were dying.

They were dying unnaturally, inexplicably. They were dying in a series of events and mishaps that seemed to some people in McFarland to resemble a curse.

"The number of deaths we've had in this community is just humongous," said McFarland police chief Vito Giuntoli. "A lot of people would call it bad luck, but I think the course of events is just plain eerie. These are kids who used to sit on my lap."

On the previous Saturday night, February 15, 1987, the day after Valentine's Day, six teenagers from McFarland and nearby Delano were killed in a head-on collision between a Chevrolet Malibu and a pickup truck.

The wreck on the highway had been an awful, avoidable, pointless episode in the midst of a tragic season.

Some months earlier, two girls from the McFarland High School track team had been hit and killed by a truck while they were out on a training run. The school's popular football coach died of a heart attack at the age of 43. Another student had drowned during the previous summer vacation.

"Logic tells us that all of these bad events over the last year are not connected," said Betty Wickersham, a McFarland High School counselor. "We have to focus on that."

The unrelated accidents, culminating in the day-after-St. Valentine's-Day deaths, elevated parental anxieties to a state of frenzy because they seemed to highlight another mounting concern. Over the past four years, McFarland's children had been suffering from a remarkable constellation of serious illnesses. Most alarming was the high rate of childhood cancers—already more than 300 percent above the normal rate expected for a town of 6,200.

The children with cancer ranged in age from a 3½-year-old toddler to a teenage football player. No doctor, scientist, or public health official could explain why so many had taken ill. Neither could anybody affirm with absolute certainty that McFarland's stricken families weren't simply experiencing some of life's unpredictable misfortunes—albeit in unnerving proximity to one another. Given the 260,000,000 people living in the United States, it was entirely possible that a lethal, but otherwise meaningless, concentration of childhood cancers could have gathered in McFarland by sheer chance. The laws of probability function with brutal impartiality; the children's tragic deaths might indicate nothing more than a statistical blip.

But in McFarland during the mid-1980s, the relevant statistics weren't available to make a sound judgment about the town's condition.

When one worried mother contacted the Kern County Health Department for information on the cancer rate in the McFarland area, she was told that there was no system for reporting and tracking cancers, except in Los Angeles and the San Francisco Bay Area. The lack of local data left open the question as to whether McFarland's cancer rate really *was* elevated above the regional norm. The only tallies available on McFarland's cancers came from hushed confidences shared between parents, and then passed along in neighborhood gossip.

To make matters more confusing, nine different kinds of cancer had been identified among McFarland's children. They included cancers of the liver, lymph nodes, bone, eye, kidney, and rare neuroblastomas affecting the adrenal glands. Most doctors believed that if there was "something out there" causing the illnesses, it should have spawned a cluster composed of a single type of cancer—but not *nine* varieties. Local public health officials were baffled.

By the winter of 1987, it appeared that McFarland might be home to an unsolved "medical mystery." Yet no medical detectives were assigned to the case. Like most rural communities, Kern County's small, understaffed public health agency had little experience tracking chronic disease. The county didn't even employ an epidemiologist, the medical sleuth who investigates these mysteries. The best that the county could offer was one nurse who had once taken a college course in epidemiology—and she was more typically involved in monitoring infectious childhood diseases, like mumps or measles, or ferreting out some local restaurant spreading salmonella food poisoning.

Many of McFarland's more skeptical residents had already decided for themselves that any health problems affecting the town were probably caused by the afflicted families' idiosyncratic life-styles. They pointed to exotic and no doubt apocryphal meals enjoyed by some of their neighbors—such as "cow's head soup." Or they accurately noted that herbs commonly imported from Mexico sometimes contained traces of arsenic. At one point, somebody even sparked wild talk about "Martian viruses" left by an undetected meteorite that might have fallen nearby. But very few people wanted to play with the idea that something might still be lurking within the town itself that was responsible for the cancers—something that could still strike any child.

Tina Bravo didn't pay much attention to the conjecture of her neighbors. Nor did she place much stock in the uncertain prognostications of the community's health officials. But the cancers winding their way through her neighborhood had begun to frighten Bravo. Something was killing McFarland's children, although the town's parents couldn't see, touch, or name it. And they certainly couldn't offer their kids any protection from it.

■   ■   ■

Standing on her porch in February 1987, Tina Bravo could see and point to the homes of a half-dozen neighbors whose children had been diagnosed with cancer.

Across the street lived the Buentellos. In August 1983, their 3½-year-old daughter, Tresa, was diagnosed with neuroblastoma, a rare form of cancer affecting the adrenal glands.

"The whole neighborhood was stunned," remembered another neighbor, Connie Rosales, who lived three houses down from Tina Bravo. "Everybody's heart just bled for this young couple. They were high school sweethearts, they'd just bought their first home, it was their first child."

Two weeks after learning about Tresa Buentello's illness, Connie Rosales' 14-year-old son, Randy, was diagnosed with cancer of the lymph glands. Randy Rosales began chemotherapy treatments; he lost his hair, and became sick, tired, depressed.

McFarland, California.

"There was something out there," Connie Rosales told her friend, Tina Bravo. "It wasn't just life, it wasn't the way it goes, the breaks."

Rosales initially suspected that the town's drinking water had made her son ill. On the flip side of her monthly water bill, she had read an oblique warning from the McFarland Mutual Water Company regarding "infants under the age of six months." The notice stated that bottled water should be substituted to feed infants since the town's water nitrate levels "may at times" exceed the State of California's drinking water standards. Nitrates probably appeared at increased levels in McFarland's water because of the fertilizers used in the surrounding fields.

The McFarland Mutual Water Company also cautioned residents not to boil the water. This practice would only concentrate the nitrates and increase the hazard. The warning was particularly important since many people in McFarland had immigrated from Mexico. They were used to boiling water to eradicate bacteria and kill parasites. They associated boiling with safety.

Like Tina Bravo, Teresa Buentello, and most other McFarland mothers, Rosales had no formal education beyond high school. She knew nothing about physiology, biochemistry, toxicology, or public health. But her son's doctor had told her about "methemoglobinemia," a disease affecting babies that prevents their blood from carrying oxygen. Nitrates in the drinking water could cause methemoglobinemia. And nitrates could also convert to nitrites and nitritosamines— and pose a risk of cancer.

Rosales stormed around her neighborhood, spreading the news about her "discoveries." She told Tina Bravo that she wanted tests performed on the drinking water by independent researchers to determine if any evidence of a health threat could be found. When she confronted the McFarland Mutual Water Company with her nitrates theory, the local manager instantly dismissed Rosales' ideas as nonsense. He told Rosales that she was behaving recklessly by alarming people about the water's safety without proper evidence. Her actions might panic the community.

"The water company told me to shut my mouth or I was going to get sued for everything that I had," said Rosales. "It was getting around town and people were beginning to think maybe there was something wrong with the water. The water company did a good job on me because I was intimidated. I was truthfully scared. This house for my children is the only thing that I have in life, and I could have lost everything."

For the moment, Rosales acceded to a truce with the water company. But soon news spread about another young man, two blocks away from Tina Bravo's home, who had testicular cancer. And the neighbors began to talk about Kiley Price, who had been diagnosed in February 1982 with Wilm's tumor, a tumor of the kidney. Eventually Kiley Price's tumor would consume 90 percent of her right kidney.

Within weeks, Sally Gonzales, another McFarland mother, telephoned Rosales and asked her if she had a son with cancer. "Someone told me to call you," said Gonzales, "because *my* 9-year-old son, Franky, just had his leg amputated." Sally Gonzales needed to talk with another mother who could understand her ordeal. Franky had osteogenic sarcoma, a form of bone cancer.

The quiet streets surrounding the block where Rosales and Bravo lived began to take on a distressing new aspect.

"We had nothing but little kids running around the neighborhood with bald heads and dark circles under their eyes," remembered Rosales. "They had that weird chemo color."

"There were three kids with cancer within two blocks in two months," recalled Teresa Buentello. "And a month later another little girl was diagnosed. And it kept going on and on and on."

During the hot Central Valley summer of 1984, Tresa Buentello, then 4½ years old, died when her tumor spread through her lung and caused it to collapse. She was the first of the children with cancer to die.

Tresa Buentello's chemotherapy treatment had totaled $500,000 at Children's Hospital in Los Angeles. While in Los Angeles, the Buentello family stayed for five dollars per night at the Ronald McDonald house, which had been established to aid families with children suffering from cancer. In the future, other McFarland families accompanying their sick children would follow their example.

"She died in my arms in the hospital," remembered Teresa Buentello, who was 27 years old at the time. "I was holding her in my arms and I had my hand on her heart and I felt her last heartbeat. And then, for just a split second, I felt that I wasn't there either—that I had died, too."

In February 1985, Adrian Esparza, 7 years old, was diagnosed with rhabdomyosarcoma, a muscle cancer that formed a massive tumor behind his eye. The Esparza family lived three houses away from Tina Bravo. Down the street, Randy Rosales was still undergoing chemotherapy.

"As soon as they said 'tumor,' I immediately thought of all those other children in town," said Adrian's mother, Rose Mary Esparza. "I was scared. I panicked. I knew Franky Gonzales, who had his leg cut off because of cancer. And there was Randy Rosales, Connie's son. I knew there was a problem here with cancer. But you really don't pay attention to stuff until it hits home. And then they said my son had a tumor."

In February 1986, Carlos Sanchez, who lived three blocks from Tina Bravo, was diagnosed with lymphoma. He was 20 years old.

One year later, on March 16, 1986, Franky Gonzales died. The effort to stop the spread of the disease by amputating his leg had failed. His parents were overwhelmed with remorse.

"How do you tell a kid that's wild and active that he's going to have his leg cut off?" asked Franky's father, Borjas Gonzales, after his son's death. "But the cancer was moving really fast, and they said that if they removed his leg it would keep it from spreading."

"After the surgery, they said he'd be okay," remembered his mother, Sally Gonzales. "But eventually it spread through his lungs and they kept giving him chemo and after the third session he didn't want to do it anymore. He was just too tired to keep fighting. And he was in pain and sick from the drugs. He was really scared and he said he couldn't do it anymore. He didn't know he would die even then."

"That morning he kept saying, 'Poppy, Poppy,'" remembered Borjas Gonzales. "That's all he kept saying. Then he quit breathing."

- - -

By 1986, life in McFarland had utterly changed for Tina Bravo. She was 31 years old, with three children at home, and she had begun to fear in earnest for their health. Mario, 14, was her oldest. Her daughter, Yadira, was 8 years old. And there was a new baby, Eduardo. Each night she prayed for their safety.

Yet when Bravo now talked to some people in town about the puzzling health problems affecting McFarland's children, she found that the fear gripping many young mothers and fathers quickly flared into anger. People had grown tired of hearing about a problem that had no probable cause—and thus, no possible solution. Some people began to whisper among themselves that Connie Rosales was exaggerating or crazy or simply clamoring for attention. Bravo knew better. Rosales was brash, loud, uncompromising—but she wasn't crazy or keen on public notice. Bravo herself could see that McFarland seemed to be suffused in tragedy.

Bravo's 7-year-old niece, Mayra Sanchez, had just been diagnosed with a brain tumor.

The first signs of Mayra's illness were the severe headaches that sent her home from school crying. "I knew that I had to take her to the doctor," remembered her mother, Esmeralda Sanchez. "We have so much cancer here." But Mayra's cancer wasn't diagnosed until the family was visiting relatives in Mexico; then her headaches suddenly exploded into wracking pain, and Mayra had to immediately seek treatment at a Tijuana hospital. After returning home, she spent a month in Los Angeles where specialists confirmed the bad news. Mayra's parents remained quiet about her illness, sharing their sorrow exclusively with close family members.

Immediately after her operation for the tumor, Mayra was unable to move at all. Eventually she managed to open and shut her mouth, shift her legs and arms, and respond to the touch of her parents. Mayra rolled her eyes up to gaze at her mother and father when they stroked her cheek.

Tina Bravo's 14-year-old son, Mario, was spending much of his free time after school with Mayra. Mario seemed particularly attached to his younger cousin, almost like a brother. He brought her stuffed animals and quietly talked with her, trying to lessen her fear.

Mario Bravo's attention to his sick cousin earned him respect from his family. Despite the mounting worries of his mother and aunt, Mario did not seem to be deeply disturbed by his intimate knowledge of cancer. "He was curious about things," explained Tina Bravo. "He wanted to know what a tumor was, where it was, what causes it. He wanted to know what happened to Mayra."

Although Tina Bravo began to worry about continuing to reside in McFarland once Mayra's cancer was diagnosed, her son still enjoyed his life in the small farming community. Mario was a popular boy, well-known throughout the neighborhood. "If he saw something in somebody's yard," said Bravo, "he'd go up to them and start asking questions, nosing around." At school, Mario continued to play the class clown. "In fourth grade, the teacher didn't know where to put him. She had him in all four corners of the classroom, but nothing worked. He was a real joker, he just had the knack for making people laugh." Mario was also mildly dyslexic, reversing letters and numerals. But he worked hard to learn, responding avidly to encouragement by his teachers. Other children seemed to be drawn to him; he typically emerged as the leader whenever a group of kids gathered.

"Whatever ideas he had, the other kids wanted to follow along," said Tina Bravo. "He would make a little ramp for his bike on the corner. And pretty soon you'd see all the kids making ramps on their corners because they had seen Mario do it. When the boys were playing football, Mario seemed more like the coach than the coach did."

As Mayra Sanchez grew increasingly ill, Tina Bravo and Mario heard more stories regarding other sick children in the surrounding communities.

Five-year-old Johnny Perez Rodriguez, of nearby Delano, had died from a neuroblastoma following months of chemotherapy and radiation treatment. Both of his parents had been employed as field laborers where they, like most farm workers, worked with pesticides and other chemicals. His mother had worked in the fields until she was eight months pregnant with Johnny.

At his funeral in Delano, Johnny Perez Rodriguez tumbled out of his small white coffin while being lowered into the ground. He spilled facedown into an open grave. His father jumped into the grave, snatched up his son's body, and returned it to the coffin. The crowd of relatives and close friends were hysterical. Earlier in the day, cemetery workers had originally buried the boy in the wrong gravesite. They had to dig him up to place him in the correct location.

Tina Bravo talked with mothers in her neighborhood about the boy's death. At church, they prayed for his family. But Johnny Perez Rodriguez would not be included among the dying children when local health officials finally began their official body count. He did not live within McFarland's city limits.

Three weeks later in McFarland, on August 30, 1987, Tina Bravo's niece, Mayra Sanchez, died of her brain tumor at the age of seven.

She was buried on September 1, 1987.

Although Mario Bravo had grown very close to his cousin, he did not attend the funeral. Instead his mother drove him to the hospital where he was found to have a hepatoblastoma, a rare liver cancer.

On the previous day, the family had taken a trip to the beach where Mario complained about "something inside" that made him uncomfortable when he was lying on the sand on his stomach. When he rolled over on his back, his mother discovered a lump below his rib cage.

"I wasn't even at the funeral for Mayra because I was at the hospital with my own son's cancer," said Tina Bravo. "I look back at that now and I know I was in shock. I couldn't reason anything out. I couldn't even think of what had happened to my sister-in-law—the nightmare they had gone through. And now it had started with my own son."

Mario Bravo was a close friend of Adrian Esparza, already being treated for cancer. He played on the same baseball team with Randy Rosales. His family lived across the street from the Buentellos, who moved to Los Angeles after Tresa died. "But Mario didn't talk about being afraid," said Tina Bravo. "Like any of us here, he didn't think it was going to happen to him."

Day after day, Mayra's mother visited Mario, helping Tina Bravo despite the recent death of her own daughter. "There was nothing they could do but try to control his pain," remembered Tina Bravo. "The pain was very severe when the tumor got big. They told me it was a rare tumor, hard to deal with in adults, and even much more so in children. The cancer had already spread to his bones."

Word about Mario's illness traveled throughout town. Teachers, friends, even total strangers, flocked to his house.

"I had a lady come here one night about 11 o'clock," remembered Tina Bravo, "when we were already in bed. It hurt her very much to know about Mario, and she wanted to do something for him. I said, 'I'm sorry, lady. I don't even know you. And everybody's asleep.' She came back the next day with a couple of balloons and a little gift for Mario and asked him what it was that he really wanted. Something that he could use right now. He had always wanted a Nintendo system. She said, 'OK, I'll get it for you.' Apparently she went around and got money and she bought it for him. I don't even know who she is."

At the homecoming game at the end of McFarland High School's football season, several families set up a booth, with Mario's picture tacked to the front, selling corn on the cob. The proceeds helped the Bravos pay for the expensive treatment their son was receiving at a clinic in Los Angeles. Other families organized a raffle with prizes donated by local friends and merchants. Several crews of fruit pickers, seasonally employed in McFarland's fields, set aside money so the Bravo family could pay its bills. Mario's mother had stopped working to spend more time with him. Before Mario was diagnosed, Tina Bravo and her husband both had worked in the fields picking grapes.

Mario spent two weeks in Los Angeles at the UCLA hospital. He returned home to a surprise party. But it was not a joyful occasion.

"I knew he was dying," said Tina Bravo. "I wanted him to die here at home around his family and friends."

His mother finally brought him to the nearby Delano Community Hospital. Tina Bravo didn't get any sleep for the first two nights as she clung to his bedside. "If he didn't hear my voice, that's when he'd say, 'Where's my mom?' " One of the nurses finally opened up the room next door so that Tina Bravo could take short naps. Her elder sister also came to the hospital; together, they stayed another two nights.

Mario's doctor attached a morphine drip. The boy's blood pressure dropped, his breathing lagged. "With the last couple of breaths, it seemed like he had fallen into a deep sleep," remembered Tina Bravo. "Then all of a sudden, he would gasp. I was praying that God would take him. He seemed to be struggling."

Three months after his cancer was diagnosed, on Thanksgiving Day, 1987, Mario Bravo, aged 14, died of cancer.

His death, coming directly upon the heels of his cousin's—and so many other children's—sent the town reeling.

"Mario's death woke up the town," said Tina Bravo. On the day he died, several Southern California television stations sent reporters to McFarland to cover the story. "I wanted people to know what happened to him," said Tina Bravo. "I thought that way we might get some answers. Expose everything."

Nearly two hundred McFarland and Delano residents turned out for Mario's funeral. A McFarland mother named Martha Salinas asked permission to form a procession from the mortuary to the cemetery. "That was something people in town wanted to do for Mario," said Bravo. She sanctioned the march as an honor to her son.

The United Farm Workers joined the march. Since the early 1960s, the UFW had questioned the safety of exposure to pesticides. Spreading the news about pesticide dangers had recently emerged as a key tactic in the union's organizing and fund-raising strategy. In the eyes of the UFW, Mario Bravo—and the other children who had died in McFarland—were all victims of pesticide exposure from the nearby fields.

At the mortuary, UFW supporters unfurled their union flag—and then the trouble began. The bold red flag with its black Aztec eagle in a white circle had long been controversial in the Central Valley. During the 1960s grape strike, the growers used to call it "Chavez's Trotsky flag." Even UFW members were initially unnerved by its powerful image. When the flag was first displayed to the fledgling union membership in 1962, some workers complained that it looked like a Communist flag, others that it resembled a Nazi banner. "It's what you want to see in it," Chavez told them, "what you're conditioned to. To me it looks like a strong, beautiful sign of hope."

But the people of McFarland weren't feeling hopeful on the occasion of Mario Bravo's funeral. And although Tina Bravo had formerly been a UFW member, and she agreed with the union that pesticides presented a grave danger, she also felt enraged that her son had been transformed by his death into a political symbol.

"When I came inside the mortuary," remembered Bravo, "I asked that whoever was in charge there to please tell them to put their flags away." The union leaders promptly agreed. But when UFW vice-president Dolores Huerta arrived, she tried to convince Bravo to endorse their message. Huerta argued that she was also a mother, that she understood what Bravo was going through.

"That got me *really* mad," recalled Bravo. "I yelled at her. 'No. This is *personal.* If you want to march with the people who are marching, go ahead and do it. But not with your flags.' "

After the funeral ended and people returned to their homes, and the UFW flags were set aside along with the festering political arguments and neighborhood grudges, Mario Bravo's death assumed even greater importance. It proved to be the turning point in McFarland—the moment when the isolated illnesses of many children suddenly welled up in the public imagination as a collective menace, assuming an unprecedented urgency and sense of violation. The community convulsed in agony, united for the moment as one body.

"When Mario was first diagnosed," said Tina Bravo, "we were really shocked and scared. We wanted answers. But people said that we must be crazy, that we were never going to get any. After he died, there was talk all over town. Everybody was asking who was right and who was wrong?"

In McFarland most people now finally agreed that there was a problem. Somebody, reasoned Tina Bravo, should be able to tell them what the problem was—and how to stop it from taking the lives of any more children.

Poisonous Doubts

Gilroy, California. Eleven-year-old farm worker, Alejandra
Sanchez.

As we said goodbye at the front
door of Tina Bravo's house, she urged us to visit Delano, just five miles
north of McFarland, where she had been raised. In Delano, said
Bravo, there was a boy named Felipe Franco who had been born with
tiny flippers where his arms and legs should have been. In her mind,
the boy's birth defects indicated that the entire region might now be
under seige.

We drove to Delano and found Felipe Franco's mother, Ramona
Franco, at her own mother's house. As we awkwardly tried to explain
the reason for our visit, the young mother smiled tolerantly and dismissed our apologetic stammers with a wave of her hands. She invited

us inside, and we followed her into the kitchen where a small army of children milled around the room as it filled with the scent of beans, onions, garlic, and homemade tortillas, all combined in a steamy *sopa*.

Felipe would be home soon. "He's out running around with the other children, and he's always the last to get here for dinner." Franco stole a quick glance at our faces, toying with our expectations of her deformed son. "I no longer think of what happened as a tragedy," she explained. "Felipe is like every other child now. He drives his electric wheelchair by pushing a control lever with his shoulder and goes all over the place."

Felipe's grandmother turned from the stove where she energetically stirred the pot of soup. "And don't forget to tell them how popular he is," she instructed her daughter. And then with the stern, emphatic courtesy of the Latina matriarch, she spoke directly to us: "*Siéntense, siéntense.*"

We sat down around the kitchen table and joined the family for dinner. After a few minutes, Felipe eased into the room on his electric wheelchair. Leather straps held his torso to the seat and back of the chair. His grandmother ladled up servings of beans and soup for everybody as Ramona Franco told us her story.

"Always, we notice the powder of the pesticides on the leaves," she said. "I worked in the fields from the beginning of my pregnancy. I was still groggy from the Caesarean when the doctor told me, 'I have bad news.' He simply said Felipe was born without arms or legs."

Franco surveyed the table to make certain that all the children were eating and that somebody was helping Felipe. Gulping down the beans spooned up to his mouth by his older sister, he teased the other kids who wanted to ride with him on the back of his wheelchair after dinner. His mother shushed him, ordered him to eat, and then go do his homework.

The fields where Franco picked grapes during her pregnancy had been sprayed with numerous pesticides, including Captan—a fungicide whose chemical composition resembles Thalidomide, the sedative that caused thousands of European children to be born without arms or legs in the 1960s. "Felipe's father was there at his birth," said Franco. "And my mother also. His father was crying, my mother even more. When we first saw him we felt such a great sadness."

No one could be certain that Felipe's birth defects had been caused by his mother's exposure to Captan. Every year, thousands of children are born with limb deformities, although they have had no contact with the pesticide. But to an increasing number of people in the Central Valley, the tragedies of Mario Bravo in McFarland and Felipe Franco in Delano did not seem isolated or unrelated. They

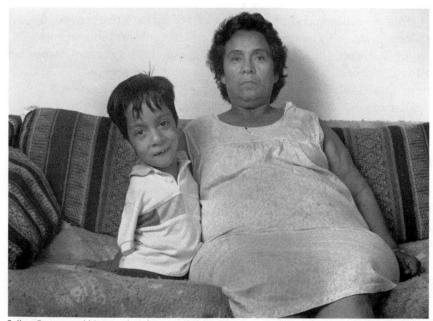

Felipe Franco and his grandmother, Maria Cervantes.

bled together, forming a pattern of risk and uncertainty that now threatened every family.

Over the next few months, we made several trips throughout California, stopping to speak with other people whose children had suddenly died or taken sick in the midst of suspicion about the contaminated environment. At the beginning of our journey, we tried to set aside our own prejudices and beliefs about the veracity of claims that linked any one illness to a specific toxic exposure. As with Felipe's birth defects and the pesticide Captan, the question of proof was enormously complicated. And as the chemical industry accurately argued, scientific studies had yet to conclusively prove that low levels of toxic chemicals could cause specific, identifiable health problems.

The question of proof also was a distraction from the complex human responses we were trying to understand. As state health officials were already complaining, the growing weight of public concern was not cast in solid reason or scientific fact; rather, it could best be measured in terms of dread. Instead of immediately burrowing into technical reports and scientific papers, we wanted to talk with people around the state to learn how their dread had accumulated—how they had come to believe that toxic chemicals were directly responsible for the pain in their lives and the damage to their communities. We were curious about the source of increasingly commonplace fears and convictions about toxic chemicals, and the ways in which the abstraction of "the environmental crisis" had been translated into a palpable menace in the lives of ordinary people.

In the small town of Fowler, California, we met Danny Shepherd, the local barber, who explained to us how his views on the environment had been distinctly altered one summer day.

Shepherd was closing up his barbershop at the end of the day's work. His 4½-year-old daughter, Jennifer, had spent the long, hot afternoon alongside him. It seemed like an ordinary day until they began to walk home and Shepherd grew annoyed with Jennifer's faltering pace.

"She kept lagging way, way behind," he recalled, "and I kept yelling at her. 'Come on, get up here, stay with me.' Usually this kid's running a block ahead of me, but this night I had to push her along. She kept saying, 'Daddy, I can't make it, I can't make it.' I had both arms full of groceries so I couldn't pick her up. We went into the house, and Jennie just lied down and her lips turned blue and she was really pale and having difficulty breathing."

Danny Shepherd and his wife, Pat, immediately brought their daughter to the hospital. The doctor scheduled a series of blood tests and promised to call the Shepherds that night with the results.

"I was in the shower when he called," Shepherd told us. "Pat heard it first. She got me out of the shower and said, 'You got to talk to him. *Now.*' He said Jennie had leukemia. My knees buckled and I hit the floor. What that meant to me was death. Our baby was going to die."

"You see," interrupted Pat Shepherd, "we had already watched one boy in town, Chris Guerra, die of leukemia. He lived eight blocks from us. And there was Fred Harlan, who went to the high school down the block—he had leukemia, too."

With Jennifer's leukemia added to the others in town, the rate of the childhood blood cancer in Fowler was 350 percent above the national average. To Danny and Pat Shepherd, the reason seemed obvious. State and privately sponsored tests had shown that the wells supplying drinking water to the town from an underground aquifer were contaminated with potentially cancer-causing chemicals: uranium; the dry-cleaning solvent, PCE; and pesticides 1-2D and DBCP. Yet each of these chemicals was present at a concentration below the level judged dangerous by health officials. "The city and health department had a meeting," remembered Danny. "They told us that to the best of their knowledge our water was safe to drink. The chemicals were at 'safe levels.' But we know there's no health data on the mixture of DBCP with uranium and those other chemicals in our water. We do know that Chris, Fred, and Jennie were diagnosed with leukemia within three months of each other.''

To the Shepherds, and the parents of other stricken children in Fowler, the evidence against the chemicals stacked up conclusively. But to public health officials, who could not dismiss the fact that leukemia cases occur even in towns with the purest water, the case against the contaminated wells seemed merely circumstantial. And there were also practical and political realities to consider. Potentially cancer-causing chemicals had already seeped into almost one-third of the state's drinking water wells at levels deemed safe. If these wells were shut down without irrefutable evidence of their danger, countless small specks of towns, like Fowler, would quickly dry up and blow away. And in the thirsty region of California's Central Valley, gripped by a multiple-year drought, the quest for alternative water supplies might prove economically unfeasible and politically disastrous.

At the roots of this conflict lies the simple, but distressing fact that little was known about the effects of mixing trace amounts of chemicals in countless combinations in thousands of water supplies. As a result, people's concerns were predictably met with bureaucratic inaction and frustrating proposals for future studies. For people like Danny and Pat Shepherd, now forced to trade their life of everyday certainties for a world of doubt, this simply would not do.

Danny's response to the questionable association between his town's contaminated water and his daughter's illness was direct: ''I wanted to blow that well sky-high,'' he raged.

Yet somewhere between blowing up the water supply and lapsing into bureaucratic paralysis, there had to be a more efficacious plan.

■   ■   ■

By the end of the 1980s, a pockmarked pattern of environmental mishaps and inexplicable illnesses had emerged throughout California. And yet the vast majority of citizens presumed themselves safe, at least for the present—or until scientific evidence could be marshalled to outline the exact parameters of risk. The health problems of small towns like McFarland, Delano, and Fowler were largely perceived as being too peculiar or remote to muster general concern. Taken individually, each community seemed to bear as little relation to the state's overall health and prosperity as the eruption of some volcano on a faraway South Seas island. If you can't see, smell, taste, or feel it in your own home, it's not a problem.

At times, the power of denial enabled people to respond to the most blatant risks with a reluctance that later would seem cavalier.

Jennifer Shepherd, the sole survivor among the children who developed leukemia in Fowler, California. The well behind her was contaminated with pesticides and dry-cleaning solvent.

In Bakersfield, about 90 miles south of Fowler, at the southern edge of the San Joaquin Valley, the pesticide DBCP had worked its way from the contaminated groundwater into the drinking fountains of Norris Elementary School. Although school officials knew about the contamination, they held off notifying parents for four years. The officials later explained that they were trying to avoid a "panic." DBCP is a suspected carcinogen; it also decreases human sperm counts, causing sterility. When the news finally broke, Norris School parents and administrators found themselves saddled with an awkward dilemma having as much to do with maintaining small town morality as protecting the public health. For months, the school district stalled action as administrators debated whether they should collect semen samples in order to test for lowered sperm counts from 12-, 13-, and 14-year-old boys who had once attended the school.

Other feats of denial seemed even more reckless than the school administrators' four-year lag in telling parents about the contaminated water.

In Santa Monica, there is a familiar pier that juts out about one hundred yards from the sandy white Southern California shore, slicing across the blue bay. Along the beach, hundreds of lithe, tanned bodies lounge under the sun, as Los Angeles rises up upon the smoggy horizon. A grand, stately azure-blue art deco archway crowns the entrance to the pier, proclaiming: "Santa Monica Yacht Harbor, Sport Fishing, Boating, Cafes." On the metal railings that run along the edge of the pier, the county's health officials have fastened more ominous metal signs stamped out in Spanish and English: "*Aviso Relacionada Con El Pescado Deportivo*, Warning Concerning Sport Fishing." The signs inform those fishing about the contamination of the bay by dangerous levels of DDT and PCBs. They specifically advise against eating the bay's plentiful white croakers. The croakers habitually feed in an area filled with several thousand metric tons of pesticide- laced sludge dumped over the years into the outer limits of the harbor.

We strolled out onto the pier to talk with the morning fishermen who had gathered in clots along the railing, struggling with their hooks and tackle. Many of the young men were recent immigrants from Laos, Vietnam, China, Cambodia, Thailand, and Korea; they could no more answer our questions than they could read the warning signs positioned exactly below the iron bars where they now lodged their fishing poles. Others read the English or the Spanish signs, but they continued to pull their day's catch out of the contaminated bay to feed hungry mouths at home; they would take their chances. Many Latin American immigrants who fished from the pier quietly mocked the Spanish warning signs. They described stinking rivers of human waste and rippling jet-black ribbons of industrial effluence that had poisoned their waterways in Mexico, Guatemala, El Salvador, Colombia, and Peru. They could neither see nor smell the problem in the roaring waves of their new home. "The fish tastes good," one man explained to us in Spanish, "my health is good. That sign means nothing to me."

Throughout the 1980s, the nation's response to the burgeoning toxic threat seemed to reflect the confused state of Santa Monica's polluted bay. A potential danger to public health might be lurking and some signs were even posted. But many people weren't reading the signs, or they weren't reading them correctly, and hardly anybody knew how to respond to their ambiguous warnings.

As we traveled throughout the state, talking with dozens of parents whose stories recalled Tina Bravo or Danny Shepherd or Ramona Franco, we soon realized that people's worst fears were almost always compounded by their own struggles with uncertainty. Seldom were the causes of an illness readily identifiable, nor the results of a toxic contamination predictable; every tragedy had its twin in doubt. The lingering questions of cause and effect were suffused in complexity.

It's not surprising that most people cannot tell polychlorinated biphenyls from trichloroethylene—or distinguish the peculiar health problems of McFarland, Fowler, or the Santa Monica pier from the uncertain threat of the latest toxic disaster to seep onto the back pages of their own hometown newspaper. And yet within the small towns where a toxic mishap suddenly appeared or kids grew mysteriously ill, people's view of the world began to undergo a profound change.

In a sense, the distraught parents whose worlds had suddenly collapsed under the weight of their children's inexplicable illnesses were only experiencing with far greater intensity what millions of other Americans had begun to feel in a more vague and inchoate manner. Fear of toxic chemicals was bearing down heavily upon the nation.

In 1989, following a report by the National Resources Defense Council predicting that six thousand children might eventually get cancer from pesticide-laden fruits and vegetables, *Newsweek* conducted a poll that showed 38 percent of the American public "worried that the food they eat may be contaminated." Every day the media seemed to serve up more bad news regarding previously unimagined dangers—the Alar in our apples, the radon beneath our homes. In 1989, the *New York Times* ran over fifty stories about hazardous substances polluting the environment; the *Los Angeles Times* published more than two hundred articles about environmental threats to the public health. By the beginning of the next decade, similar stories would appear almost daily in papers throughout the country—a national police blotter of environmental suspicions and chemical criminality.

Almost all serious discussion of the relationship between toxic chemicals and human illness seemed to inevitably end on an ambiguous note. Even the most expert scientific judgments regarding environmental health hazards were bound up in loose threads; their conclusions read like maddening jumbles of imprecision and doubt. In 1982 the pervasive dioxin contamination of Times Beach, Missouri, led to the evacuation of the entire community, making the town a national symbol of toxic disaster, second only to Love Canal. Yet nine years later, Dr. Vernon N. Houk, the federal official responsible for the evacuation, announced that his decision had been a mistake, based on faulty data subsequently discredited in the late 1980s by scientists who energetically downgraded the health peril of dioxin to human beings. Dr. Houk's pronouncement on dioxin's safety was greeted rapturously by the chemical industry as a symbolic recantation of the government's regulatory overzealousness. Five months after Houk's public announcement ("I would not be concerned about the levels of dioxin at Times Beach," he assured the public), German researchers published startling new findings on dioxin that indicated even greater cancer risks to human health than had been previously assumed. "I hope that this study would give them pause," said an author of the German study, referring to plans to raise the level of acceptable dioxin intake for humans. For the former residents of Times Beach who had lost their entire town in 1982, the science of 1991 was like watching a dizzying Ping-Pong game.

In truth, the experts in government, science, and industry knew almost as little as the general public about the long-term health effects produced by the majority of 65,000 synthetic chemicals now used throughout the world. According to the National Academy of Sciences, most Americans are exposed daily to tens of thousands of chemicals that are "legitimate candidates for toxicity testing related to a variety of health effects," but "only a few have been subjected to extensive testing, and most have been scarcely tested at all." The Congressional Office of Technology Assessment has estimated that between 80 and 90 percent of all chemical products have not been tested for their impact on human health. These analytic loose ends inevitably undermined the oversight and regulation of potentially dangerous chemicals. A California state law passed in 1984 that required manufacturers to submit plans by 1987 for the testing of two hundred chemicals presumed to be most dangerous to human life resulted in only one significant action: the postponement of the deadline for four more years. To this date, the companies have yet to comply.

Today, uncertainty remains the rule in calculating the amount of disease, if any, caused by our exposure to toxic substances. And as a result, people may either regard the lack of hard data as a rational justification for ignoring the problem; or, alternately, they may presume without evidence that the contaminated environment is the source of their physical ailments.

Of course, our national anxiety level did not rise overnight. It has been steadily mounting for three decades.

In 1962, Rachel Carson's *Silent Spring* set the tone for contemporary fears when she decried "the chain of evil" through which DDT and other "sinister" chemicals might eventually alter "the very nature of the world—the very nature of its life." Suddenly, from all directions, North Americans seemed to inhabit a subversive environment mismanaged by the faltering magic of unreliable technocrats. The water that sustains us, the air we breathe, the ground below us might one day without warning rear up as a hideous head of poison—a massive, unanticipated by-product of the brave new chemical complex that has spread throughout the environment since World War II.

The force of these doomsday hypotheses, combined with a lack of basic information on exposure and toxicity, has reinforced the sense of danger that many people have begun to feel. And psychologists studying the nation's contaminated communities agree that "the worst kind of threat is dread of the unknown." For a growing number of North Americans, their exposure to often invisible contaminants equals a perplexing, terrifying, and perhaps permanent immersion into a dark, dreadful ocean of the unknowable. Residents of toxic towns have found themselves wondering whether minor health problems might someday erupt into major illnesses. Their minds turn to the unanswerable questions: Was our town really contaminated? How severe was the exposure? Has it stopped? Will it return? Was I harmed? How badly? Will I become sick in the future? Will it affect my children?

"Exposure to toxic materials not only changes what people do, it also profoundly affects how they think about themselves, their families, and their world," explained Michael R. Edelstein, a psychologist who studied the emotional and social impact of chemically contaminated drinking water in a New Jersey suburb. "In short, it represents a fundamental challege to prior life assumptions."

Our experiences in McFarland, Delano, Fowler, Bakersfield, and Santa Monica changed our thinking, too.

At a distance, it had been possible to read the anger and dread of afflicted families as an understandable response to life's inevitable tragedies. Or on the far end of the scale, their reactions might even signal the paranoid's classic delusion of poisoning.

But close up, talking with dozens of people in a handful of towns, it didn't *feel* like hysteria.

Almost everybody with whom we initially talked only reinforced the pervasive sense of uncertainty that underscored the toxics dilemma. And this sense of uncertainty was not relieved when we examined the scientific and technological studies or talked with the experts.

Looking back, what seems most surprising is how long we managed to sustain our early perception of the toxics problem as a provincial oddity. We didn't yet believe that it raised important questions for the rest of us who lived outside of remote places like McFarland, Delano, and Fowler—or even less, that it somehow spoke to the fundamental condition of our country.

And then, we visited Casmalia.

-    -   -

Casmalia, California, is a sleepy, slapped-together rural fringe community banking the coast about 150 miles northwest of Los Angeles. About two hundred people live in Casmalia—most of them clustered around the town's single main street, most of them Hispanic and poor.

Twelve miles away lies Santa Maria, a booming white, middle-class suburb of the kind that has fueled central California's frantic growth for two decades. Near the coast, but distant enough from the smog, traffic, crime, and other urban indelicacies of Los Angeles, Santa Maria seemed to many residents the perfect compromise between city life and the countryside.

Drawing together these two very different communities was the 250-acre liquid toxic waste dump operated by Casmalia Resources Inc.—the source of a bilious conflict that had drawn residents in both towns onto an unexpected collision course with the State of California.

We headed to Casmalia with vague expectations of visiting a much larger version of McFarland or Fowler. And yet from the beginning, almost nothing turned out as we had first expected.

In reality, there were only two indisputable facts governing the controversy in Casmalia: The toxic waste dump operated by Casmalia Resources Inc. contained a vast quantity of chemicals deemed potentially hazardous to human health. And people in the two communities surrounding the dump—Casmalia and Santa Maria—were reporting a number of inexplicable illnesses.

Everything else—every fact, fear, theory, official declaration or public denial, every personal commitment and bureaucratic evasion—was shaded by the looming presence of uncertainty. Casmalia and Santa Maria were citadels of doubt. And the drama in which their questionable risks were being worked out drew together all levels of society, from the community's Mexican immigrants who were cast as the dump's first victims to the upper echelons of state goverment, where officials who had once been perceived as neutral guardians of the public health were soon regarded as slippery conspirators more concerned with the welfare of a conscienceless industry. In fact, the peculiarities of this toxic waste morality play convinced us that something important was happening throughout the country that transcended the boundaries of most environmental skirmishes. Casmalia signaled our first step beyond the realm of provincial, personal tragedy and into the heart of toxic America.

We met Nick Irmiter one Sunday morning just before dawn as the sun began to peek above the hills of Casmalia. Since 1984 Irmiter had lived in town with his wife and two young daughters. He had been layed off work for almost a year, but he still looked the part of an oil field roughneck—a job that he had first learned in Texas. Recently, he devoted most of his days and nights to fighting Casmalia Resources, the operators of the huge liquid toxic waste dump crowning the hills above his home. He was convinced that the dump was making his wife, kids, the entire town, desperately ill. That morning, said Irmiter, he would show us how.

Irmiter had decided the night before, over several glasses of lemonade in his home's dark, curtain-drawn living room, that at sunrise the three of us would slip under the barbed wire fence that cordoned off Casmalia Resources' land. Then he would lead us across a winding path over the hillside to view the dump before the heavy morning fog lifted and exposed us to the weekend's morning crew and their security guards. The week before, Irmiter had led a Greenpeace activist on a similar expedition across the private land to surreptitiously collect water samples near the dump for lab analysis. Irmiter claimed that the Greenpeace raider had burnt his hand in a creek turned caustic from the dump's overrun.

As the rising sun cast an orange trim along the hillside, Irmiter signaled us to follow him up the path of scrub grass and manzanita until we could reach a distant peak from which to view the evil empire. The ascent was sharp and long. We gasped for breath, until we could no longer talk comfortably among ourselves. Then Irmiter sprinted far ahead, disappearing into the fog.

For several years, Casmalia's two hundred residents had seriously wondered about the health effects of living directly downwind from the dump. These concerns did not seem surprising, given their community's unenviable proximity to the dump's 155 million gallons of hazardous industrial, agricultural, and household wastes. And yet in recent months, the most passionate opposition to the dump had not arisen from the residents of Casmalia. The most vocal and stubborn critics lived in Santa Maria, which stood 12 miles away—far from the grip of the noxious chemicals that frequently squeezed the breath out of people in Casmalia. Unlike Casmalia residents, Santa Maria's activists lacked a visceral motivation to oppose the dump. Instead, they were moved by a study. In Santa Maria, the word quickly spread that a local physician had discovered that their town—not Casmalia—suffered from an extremely high rate of birth defects.

Dr. John Barry, a Santa Maria pediatrician, had uncovered data indicating an inexplicable concentration of cleft palates among newborns. He concluded that the deformity could have been caused by industrial solvents "dumped in large quantities at Casmalia." Barry speculated that water contamination might have resulted from leakage at the Casmalia dump, though he offered no explanation of how the contaminated water could have reached the household taps of pregnant women in Santa Maria; the community's water source did not come from Casmalia. And yet Barry and 71 other local doctors took out an ad in the Santa Maria newspaper urging "the immediate closure and clean up of Casmalia Resources Toxic Dump because of water contamination, air contamination, and public health danger." The ad's message, which represented a rare display of activist solidarity among an otherwise conservative medical profession, was assembled above clip-out coupons demanding governmental action, which readers were instructed to mail to their state and federal representatives, as well as to Dr. Kenneth Kizer, the director of the State Department of Health Services. "SECOND WARNING!" the newspaper ad headline had admonished the local citizenry, "STAND UP AND PROTEST NOW!"

Department of Health Services chief Kizer must have been profoundly confused by Barry's linkage of Casmalia's dump to the cleft palates in Santa Maria. But if Kizer had to ponder the baffling assertions of local citizens, the people in Santa Maria and Casmalia were even more perplexed by the flip-flopping claims of the state. In 1987, California's Department of Health Services formally assured residents that there was no evidence of local groundwater contamination. Yet only a few months later, the State Regional Water Quality Control Board conducted tests and noted "widespread contamination" directly beneath the dump from numerous chemicals. The greatest offender was the cleaning solvent TCE, a suspected carcinogen, which tested at 6,100 parts per billion. The level at which the California Department of Health Services was supposed to take action was 4 parts per billion.

Neither the residents of Casmalia nor Santa Maria received their drinking water from the contaminated area under the dump. Nevertheless, the new report of water contamination inflamed fears about illness throughout both Santa Maria and Casmalia. Widespread common complaints, such as sore throats, headaches, nausea, respiratory problems, and chronic fatigue, were attributed to the nefarious influence of liquid toxics swirling in the waste pools amid the Casmalia hills. "I saw the same symptoms so often I gave it a name," asserted one family practitioner, "the Casmalia Syndrome." Local doctors found the Casmalia Syndrome to be prevalent in both Casmalia and Santa Maria.

In Casmalia itself, only a mile downwind from the liquid toxic waste ponds' chemical vapors, Jose Gracias lifted up his shirt to show us the multiple scars that traveled across his abdomen like railroad tracks. His belly, he explained, had been flayed numerous times by

surgeons in unsuccessful attempts to locate the source of his mysterious liver ailment. His daughter, Maria, had recently been found to
suffer from the same problem. She gazed with some intensity at her
father's scarred belly. Her own surgery was scheduled for later that
month. The Gracias family had lived in Casmalia since the dump had
opened.

For years, Casmalia residents had complained about nauseating
fumes from the dump pouring into town like "burnt hair or a woman's bad permanent." In fact, as far back as November 1984, principal
Ken McAlip had shut down the town's school and sent children home
when the acrid fumes made students and faculty too ill to continue
working.

"On the day we closed," McAlip told us, "it kept getting worse
and worse. Normally, we have gentle breezes which keep the stuff moving along. But if it stops, the fumes settle in the valley here. It was a
warm day and the wind stopped at lunchtime. The odor rolled through
the vents and into the classrooms—this metallic smell and taste. You
had a feeling of suffocating. We were nauseous, we couldn't breathe.
Then the kids who live here looked at us and said, 'Hey, this isn't bad.
You should be here at night.' " McAlip estimated that at least 80 percent of his students suffered from chronic respiratory illness.

Judging by the principal's testimony, and the compelling spectacle of the students fleeing their own school, it seemed easy to understand why Casmalia residents stood strongly against the dump. Yet the
malodorous, perhaps treacherous, fumes that descended upon the
isolated community didn't explain why such intense enthusiasm to
shut down the Casmalia dump had spread to the neighboring city of
Santa Maria. No doubt the newspaper advertisements signed by 72
Santa Maria physicians deserved some credit. It took either extraordinary courage or recklessness to ignore your own doctors' public
declarations (paid for out of their own pockets) about "highly TOXIC
chemicals that can cause cancer, leukemia (sic), birth defects, and
impotence!" But there seemed within Santa Maria a heightened commitment to close down the Casmalia dump that radiated with an intensity not usually found even among communities that suffered from a
far greater degree of demonstrable toxic exposure. More than any
community we had so far visited, Santa Maria seemed unified and
mobilized for action. The prosperous, white, middle-class suburb of
Santa Maria had somehow seen its own future in the face of rural,
poor, and largely Hispanic Casmalia.

After four frustrating years of battling the foul-smelling fumes
that floated off the dump and intermittently forced the Casmalia
school to close, principal McAlip and the residents of both Casmalia

and Santa Maria were ready for more drastic measures. On May 16, 1988, more than 80 students, parents, and teachers drove 150 miles to the State Capitol in Sacramento to stage an audacious protest. *Today, the residents and school children from the small town of Casmalia have closed our school and will hold classes on the steps of the Capitol,* read the leaflet passed out by parents participating in the highly visible media event. *What could lead reasonable people like us to such drastic actions?*

As the demonstration proceeded, this rhetorical question was quickly answered—at least on the most obvious level. Children and teachers stepped up to the makeshift podium to instruct in a scratchy microphone voice that the chemical winds wafting over the hills into Casmalia from the toxic waste dump were poisoning students in their own classroom. A young girl recited a poem that she had composed for the occasion. Its first verse began:

> Dear God who watches over me
> Please help me solve this mystery
> They're hurting my family
> They're killing my community
> Oh why God is this government doing this to me?

"This is a long way to go to get to school," declared Angie Irmiter, Nick's wife and the spokeswoman for a Casmalia parents' group called MIFT (Mothers Involved in Fighting Toxics), "but these kids have a lot to teach our legislators about what it's like to live and go to school next to a toxic waste dump. We can't afford to pay our medical bills, and we can't afford to move. Our only hope is to shut the dump down once and for all."

The crowd marched through the doors of the capitol building, then straight into the governor's office. The reception room filled with parents and children waving banners and signs. The harsh lights of TV cameras flooded the room, lighting the face of the governor's secretary as she wanly explained that despite his great interest in their problem, the governor would be unavailable to meet with them today. The secretary accepted the stacks of petitions that demanded the dump's closure, promising to pass them on to her boss. On the top of her desk, now brimming over with pamphlets and flyers, somebody placed an enormous photograph that the secretary certainly would not pass on to the governor nor anybody else. The photograph drew into sharp focus a newborn's face, deformed by a cleft lip and palate. The baby was from Santa Maria. The words underneath the portrait read: "Close the Casmalia Dump!"

Media attention at the State Capitol seemed to set the wheels of government spinning. Soon after the demonstration, officials with the State Department of Health Services agreed to meet with a trio of community representatives. Oddly enough, none of the "community representatives" actually lived in Casmalia. Patricia Prisbrey and her neighbor Kathy Hoxie were activists from Santa Maria. Principal McAlip, although he had long been the publicity point man in the battle against the dump and spent every working day in Casmalia, lived in another small town nearby. The trio from Casmalia/Santa Maria vividly recall meeting with Dr. Ken Kizer, the department's chief. Kizer denies that he attended the meeting.

Kizer had already earned the enmity of his opponents in Santa Maria and Casmalia for his public pronouncements regarding what he believed to be an hysterical response to the dump. "I think what's going on," Kizer had told reporters, "and I don't want to sound harsh, is a lot of toxic superstition. It's kind of like when a black cat walks in front of you and you have an accident. Well, the cat didn't cause the accident." When principal McAlip joined with Prisbrey and Hoxie in Sacramento, it seemed as though the relationship between the state government and the communities only had room to improve. In fact, during the single hour of their meeting, the residents' disillusionment with experts and authorities grew considerably worse.

"We walked into the meeting room," recalled Prisbrey, "and there were three lawyers sitting across from us and on the other side three other guys, brilliant specialists in their fields of chemistry, toxics, and all that." To Prisbrey and company, it felt like a setup, an orchestrated intimidation. The trio found themselves twisting their heads back and forth between comments from the lawyers and the technical experts. But they had prepared carefully for the meeting, and they insistently focused the conversation upon Dr. Barry's study documenting the doubling of cleft palates in the Santa Maria area. The trio further argued that other doctors in Santa Maria had noted increases in childhood cancers, respiratory disease, and other problems. McAlip brought up the problems that he had personally seen in Casmalia itself, claiming 14 untimely deaths in eight years among people in their forties and fifties, including three cases of a rare blood disease.

When Prisbrey, Hoxie, and McAlip demanded that the Department of Health Services director comment on the local physicians' concerns, Prisbey remembers Kizer responding with a scoff and a shrug. "Well, *your doctors*. . . ."

"Wait a minute," Prisbrey shot back. "Don't say 'your doctors' like that. You're a physician and you went to the same schools as they did. You're no different."

There was a long moment of uncomfortable silence.

It was clear that the meeting was not going well, not going at all the way anybody had planned. Prisbey remembered the Department of Health Services chief trying another tack. He attempted to explain to the trio that he alone had no power to close the dump. He relied upon a panel of experts, he needed *evidence*. He could not operate, even if he desired, on the basis of sentiment rather than science. For example, he instructed, as he pointed a finger at Kathy Hoxie sitting across the table, he couldn't arbitrarily decide that he hated the color of Hoxie's new orange-pink blouse; he couldn't assume the purely subjective stance that the blouse made him sick every time he saw it, that the blouse was in fact *toxic* to the entire community. And he certainly couldn't ban people like her from wearing that blouse.

"Kathy was ready to cry," remembered Prisbrey. "Here she was all dressed up and this asshole was laughing at her."

By the end of the meeting, relations between the community and the state had hit a new low. When Kizer later requested medical records from Casmalia and Santa Maria residents to determine whether a pattern of illness could be discerned within the communities, the residents flatly refused. Over two hundred people had joined in a lawsuit against Casmalia Resources. They claimed that surrendering their medical records might compromise their chances for legal redress. The state argued that it needed the records to tell if a genuine health problem actually existed.

Ken Hunter, the president of Casmalia Resources, thought that the impasse had less to do with an erosion of trust between the state and the communities than the tactics of the courtroom. "The residents have an economic interest in killing us," he asserted. "The whole thing is so overblown and out of proportion to the danger, it's ludicrous." When Hunter was asked about the health problems around which the lawsuit had been focused, he rolled his eyes and tapped his temple with one finger. "It's psychological."

If boggling uncertainty within the community was the "psychological" problem that Hunter had diagnosed from his lavishly appointed office one hundred miles away in Montecito, he was certainly on the right track. The information that community members had managed to pry out of Casmalia Resources, the California Department of Health Services, and the federal EPA all fed this confusion, at times raising the temperature of public discourse to a panic fever. Skeptical residents in Santa Maria and Casmalia could not forget that the health department's initial statement assuring no evidence of groundwater contamination beneath the dump had been almost immediately contradicted by the state water board, who found significant problems. As a result, many citizens were disinclined to believe later reassurances from the health department that the contamination under the dump had not and would not reach their town's water supply in the future.

Air testing proved equally controversial. State surveys failed to turn up evidence of airborne contamination settling within Casmalia's city limits. But residents insisted that the samples had been taken on days when the wind was blowing away from the town. They hired their own experts to conduct tests that soon "revealed 17 toxins on school property that match with those in storage in greatest quantity in open ponds at the dump site."

Even less agreeable than the pretzel logic of the state bureaucracy and the stone-faced tough guy denials of the dump's owners were the legalistic contortions of the EPA. In 1989, the EPA fined Casmalia Resources a record $6.2 million for storing 270,000 cubic yards of toxic waste above its regulated permit capacity. Yet within weeks, the EPA also awarded Casmalia Resources a new permit to expand its facility. An EPA spokesperson tortuously explained that the fine and the permit were "unrelated," and the government was "obligated by law" to accept the application for a new permit, "regardless of the history of the facility."

In Santa Maria and Casmalia, fears fed upon themselves, multiplying with the energy of unbridled imagination. One citizen had assembled a "chemical contamination primer," which he titled *Until Death Do Us Part . . . Man's Poisoning of the Earth*. After ominously noting that "Casmalia is derived from the Chumash Indian word Kasma'li, which means *the last. . . ,*" the primer ran through several dubious theories asserting that toxic chemicals were the "underlying cause of such new diseases as Chronic Fatigue Immune Deficiency Syndrome (sic), Lyme disease, and even AIDS." The author had filled out his primer with two dozen newspaper stories whose combined impact was meant to pry open even more questions about Casmalia's toxic dilemma. A quick survey of the headlines revealed an odd assortment of ailments: "Bacterial Infection Linked to Mysterious Pelican Deaths"; "Mysterious Illness Proves to be Chemical Sensitivity"; "Biologists Coordinating Hunt for Abalone Killer"; "Cat AIDS Expected to Rise"; and "Scientists Suspect Ocean Pollution is Killing Sea Lions" (with a handwritten annotation in the margin—"coincident with health epidemic in Santa Maria"). It was this kind of amateur theorizing that drove public health officials crazy—and yet it could only be expected when everything remained in doubt.

With uncertainty proving the rule, many people preferred not to theorize or fight, but to run. In Casmalia, homes went up for sale at rock-bottom prices. Dr. Daniel DuCoffee, a Santa Maria cardiologist and the former president of Physicians Against the Casmalia Dump, the group of 72 doctors who ran the newspaper ads, publicly declared that local physicians were "so concerned about their own families' health that ten" had "moved out of the Santa Maria Valley in the past two years" and others were planning to move. In either town, the state of mind was best summed up by the bumper sticker hanging in principal McAlip's office at Casmalia's elementary school: *I Survived the Casmalia Toxic Dump—I HOPE!*

▪   ▪   ▪

On that early Sunday morning in August, as we trudged up and down the hills above Casmalia's toxic waste dump with Nick Irmiter, we didn't speak about any of the doubts, uncertainties, excesses, or absurdities that had shaped the relationship between the two communities, the state officials, and the owners of the dump. Instead, we carefully watched our step along the rocky hillside, tried to keep up with Irmiter's fast clip, and occasionally broke into a sweaty trot or crouched in a deep rutted gully as we heard the gear-grinding lurch of another company pickup truck filled with security guards ready to slip out from behind the blanket of fog and apprehend us. Hunkered down in the weeds after one close call, foolishly feeling like boys playing soldiers, we asked Irmiter to explain how he had first become involved in the fight.

He seemed an unlikely warrior. Nick Irmiter was 40 years old, tall and lean with a pencil-thin moustache, casually dressed in worn blue jeans and a T-shirt; nobody would have spotted him for a middle-aged man with a sick wife, two kids, no job, and a stack of bills at home. But this morning, he was obviously enjoying himself, bounding up and down the hills with jackrabbit speed and agility. Out on the chase, playing the fox before the lumbering hounds, he was showing us that he could pit himself against the company vehicles whose groaning engines were growing louder in the mist as we approached the summit overlooking the dump—and he could win. There was even something disarming and likable in his nonstop rambling about the toxic waste dump, his one true subject.

"You've seen that commercial on TV about Raid," he asked us, "where they spray the insects cooped up in the corner, and they can't get away? Well, that's what it's like here with us."

Then a company truck peeked out from the fog, its driver hailed us, and Irmiter dashed down the hillside for two hundred yards, where he was swallowed up in the blanket of mist. We followed clumsily behind.

At the bottom of a hill, we caught our breath, talked about turning back, and tried to determine what to say if we were arrested. Irmiter was not worried. He assured us that for months, while trying to gather evidence of safety violations that could be filed with the EPA or state regulatory agencies, he had scaled the hills surrounding the dump several times each week. Now, perhaps twice or three times a month, he would slip up to his favorite vantage point just to keep watch on the operation.

Irmiter had become, in the words of another Casmalia resident, "a toxic crusader at-large."

After a few minutes rest, we picked ourselves up and stumbled back along a rockier, more remote trail that Irmiter quickly located, and we ascended for another 20 minutes. Finally, we broached the crest of a dry, grassy knoll directly across from the dump.

What we saw from our perch hardly looked alarming.

Rather than the ominous fount of toxic disaster, the Casmalia dump looked more like a rock quarry. A legion of busy trucks scrambled among an accumulation of pits, vats, storage tanks, and trailers. Dirt pathways were etched in harsh grooves across the rolling brown hills, and the dull roar of industry was audible even from the distance. It was certainly ugly, and given the right perspective, it was possible to conjure up some apprehension about the huge amount of toxic wastes being buried there. But nothing in the site inspired terror or awe; fear demanded a stretch of the imagination.

Irmiter could imagine the worst.

"It's unjust," the toxic crusader uttered breathlessly, gazing out over the huge scar in the hillside. "It's criminal and it's immoral."

What could lead reasonable people like Nick Irmiter to such drastic conclusions?

It was not simply the toxic threat that was transforming lives in Casmalia and Santa Maria—but beyond that, it was the widening breach between the residents and the experts who were supposed to be guarding the community's health and safety. Irmiter and others had lost faith in the experts, the authorities, the elected officials; these people no longer seemed to know or care enough to serve and protect them. Ordinary citizens had to watch out for themselves.

When Irmiter now looked at the dump, he didn't just see danger; he recognized the combined forces of chaos that ranged from the mundane, albeit maddening procession of bureaucratic deceit and arrogance to the more vaporous, if equally disturbing kind of cosmic uncertainty that was now settling upon his own life in a dozen different ways. Irmiter was like many people in Casmalia and Santa Maria—like many people throughout the nation—who believed that America no longer "worked" as it once had, though they held only tentative notions about who to blame for the lapse and how it might be repaired. All they knew was that the first place to start fixing the country had to be the toxic dump overlooking their own homes. The dump stood as a symbol for all the other environmental problems that most Americans could only perceive as abstractions. Irmiter and his friends couldn't see the depletion of the ozone layer; they couldn't hear the rain forests being toppled; they couldn't smell or taste global warming. But the dump sat squarely in the middle of their lives, palpable and threatening. Against the dump, they could act.

And yet, as even Irmiter realized, their success ultimately depended upon some form of uneasy alliance with the authorities. And at this point, the people of Casmalia and Santa Maria, and the local, state, and federal officials, could agree on nothing.

The Mothers' Crusade

Yukon, Pennsylvania.

Diana Steck did not go to jail for the spotted owl.

Throughout the 18 hours that the 35-year-old mother of two spent in Pittsburgh's rough Northside lockup, she failed to consider even for a moment the plight of the snail darter, the snow leopard, or any other endangered species. When the threatening voice of a black prostitute rang out from the darkened corridor to assure all the other working girls that she would bust the ass of the honky protestors, Diana Steck didn't try to change the subject to global warming. At 6:30 in the morning, when the jail matron finally brought the prisoners some water, Diana wanted to drink from the Styrofoam cup as much as anybody else; she resisted all temptation to lecture her keepers and cellmates on the evils of nonbiodegradable plastics.

Diana Steck was simply not that kind of environmentalist.

By her own account, Diana was an ordinary, middle-class, working American mother, who found herself catapulted into public life—and the Pittsburgh jail—because of the huge toxic waste dump that stood at the center of her small town in rural Pennsylvania. Like hundreds of other "ordinary" mothers who emerged over the past decade to lead the fight against contaminated landfills, toxic waste incinerators, and other chemical threats, Diana did not hail from the ranks of mainstream environmentalists who grew up hiking with the Sierra Club, birdwatching with the Audubon Society, or sending checks to the National Resources Defense Council. Diana entered the confrontational world of grassroots power politics solely because of her concern for the health and safety of her family.

On April 23, 1989, Diana Steck and three other residents of the small mining town of Yukon, Pennsylvania (population, 1,100), drove 30 miles through the region's green rolling hills to arrive in Pittsburgh. In the dense center of the city, the angry contingent from Yukon marched into the State Department of Environmental Resources, along with 17 citizens from other parts of Pennsylvania, and refused to leave. Steck and her friends wanted to drive home a message to the press, public, and governor: State officials wouldn't budge on closing and cleaning up Yukon's toxic waste dump; therefore, its residents wouldn't budge from the office at closing time.

The police arrived, as expected, and arrested Diana Steck, her neighbors, Dave and Carol Rupp, and a teenaged girl, also from Yukon. Although the sit-in marked the group's first collision with the law, nobody worried about jail. Their lawyer assured them that the police would simply drive the protestors to the nearest stationhouse, run them quickly through booking, and soon after, the judge would release them on their own recognizance. Diana planned to be home to cook dinner for her family.

But the police hauled the demonstrators all the way across town to the seedy Northside jail. Then the stationhouse fingerprinting machine broke down. The prisoners' processing papers mysteriously disappeared. Nobody fed them. When they were finally arraigned, the judge set their bail at $3,000—three times as high as the drug offenders with whom they had spent the night behind bars. Everybody in the Northside jail, including the prostitutes and junkies who eventually grew more sympathetic to the protestors' story, agreed the state wanted to teach them a lesson. Diana knew why: She had ceased to be a compliant, unquestioning, respectable citizen. Like her counterparts in contaminated communities throughout the country, she raged against her government and its contempt for people like herself. Her experience over the past five years had turned her into a dangerous woman.

But by the time the judge finally released her, Diana didn't look dangerous. Along with her son, Jeremy, she suffered from a number of serious health problems, whose origins she attributed to the toxic waste dump near her home. Too sick now to walk, Diana had to be carried out of her cell.

Back in Yukon, a band of supporters welcomed Diana and her jailhouse colleagues like returning heroes. Members of CRY (Concerned Residents of Yough County), the local anti-toxics organization, threw Yukon's first political prisoners a welcome-home party. Someone baked a huge sheet cake, frosted in red, white, and blue, bearing the inscription:

WE THANK YOU
THE YUKON 4

Shortly after Diana returned home and dragged herself into bed, the telephone jangled her out of her dreams. She picked up the receiver to hear the familiar voice of her elderly grandmother.

"How could you do this?" she objected. Diana's grandmother was a staunchly conservative Polish immigrant. She had just read about the arrest in the newspaper. "You didn't have to go that far," she insisted. "You didn't have to go to jail."

"Grandmother," argued Diana, "I didn't do anything wrong. They're trying to kill Jeremy and me. Don't you think that I did the right thing by trying to stop that? They're still trying to kill people over there."

"Of all my grandchildren," complained the old woman, "I would never have expected this of you."

Diana couldn't begin to explain to her grandmother how much she had really changed.

- - -

When Diana Bengel and Gary Steck married in 1977, they argued about moving back to Yukon.

Diana's husband had grown up in rural Yough County, far from the bustle and noise of big-city life. The small town of 1,100, with its ingrown population of Yugoslavian, Czech, and Ukrainian farmers, ex-miners, and retired factory workers, seemed to him an ideal place to start their family.

Diana grew up 13 miles from Yukon, in the small city of Jennette. From Jennette, once the hub of the regional glassmaking industry, her husband's home now looked like the middle of nowhere. Yukon was coal country, named for its rich vein of "black gold." In the 1950s, Yukon's mines closed, taking with them the town's bank, stores, and movie theater. Only the abandoned coke ovens, burrowed like puncture wounds into the surrounding hills, remained as evidence of the boom years. "There wasn't anything out there," Diana complained. "There wasn't even a grocery store. It was a half-hour's drive no matter where you had to go in order to do anything."

But when Gary's parents found the newlyweds "a cute little house for a dream," Diana set aside her misgivings and agreed to start out married life in Yukon. "I was a peacemaker," Diana remembered. "At that time, I wouldn't say beans to anybody." Diana Steck stood barely five feet tall, a slender and energetic woman with dirty-blonde curls dangling to her shoulders. In her blue jeans, flannel work shirt, and thick glasses, she looked both pretty and severe, like a prim country school teacher on her day off. Diana had been an only child, strictly raised by her opinionated, second-generation Polish-American mother and her steelworker father. Diana grew up thinking of her father as "a die-hard union man" and an "Archie Bunker," who insisted that she attend a Catholic high school outside of Jennette because he feared the racial strife of the 1960s boiling over into the city's integrated public school. From an early age, Diana's parents encouraged her to study hard, listen to the nuns, and follow the rules. By the time she completed nursing school and married Gary Steck, she had visited only two big cities—nearby Pittsburgh, and once, for a nursing conference, she traveled across the state to Philadelphia.

Even to herself, Diana Steck seemed a typical small-town girl. She spent little time worrying about the outside world. She limited her reading to nursing textbooks, the local newspaper, and the occasional popular novel. She didn't vote. Like most people Diana knew, her interests focused on her family and her career. "You just want to live," she explained. "You don't want anybody hassling you. You pay your taxes. You want a safe place to bring up your kids. That's it."

But Yukon proved a rough fit for Diana. Months passed before the Stecks learned the names of their closest neighbors. "I moved three times when I was growing up in Jennette," remembered Diana. "Everywhere we landed, the neighbors would bring us supper the first night. I was waiting for that to happen when we moved to Yukon. Where's the pie? Where's the soup?" Like many insular rural communities, bound together by ethnicity and the ancient tangle of family bonds, Yukon did not rally quickly to greet a newcomer in town—even if she had married a native son.

Even the local Catholic church ignored the new arrivals. For months, Diana stood alone at the end of Mass when the celebrants were instructed to turn to one another, shake hands, and offer the ritual blessing: Peace be with you. When Diana took her infant son to Mass, the reception turned even chillier. On their first visit, Diana carried Jeremy close to the altar. An usher tapped her on the shoulder and pointed her to the back of the church, banishing Diana to the crying room, even though Jeremy wasn't fussing. "I was so upset that I didn't ever want to return to that church," she recalled. "For a while, I didn't."

Diana felt even more isolated and lonely as her family battled one illness after another. Almost from birth, Jeremy suffered ear and urinary tract infections. Since coming home from the hospital, he seemed to live on antibiotics. Diana and Gary had always been vigorous and healthy, but now they also endured an alarming number of colds, flus, infections. They found blood in their urine. Diana's joints swelled painfully, and occasionally she would lose feeling in her hands. One time she almost scalded Jeremy in the bathtub because she couldn't gauge the water's temperature.

"I used to be the kind of person who could do a million things at once, get three or four hours sleep, and be just fine," she remembered. "I was always doing something. But then it got to the point where I couldn't even get out of bed in the morning. Total exhaustion and severe pain. I could feel where everything is connected."

Diana underwent a battery of blood tests at the hospital where she worked. The test results pointed to numerous abnormalities, indicating that the inflammation of her joints and damaged nerves were probably caused by her immune system kicking into hyperactivity to fight off some mysterious malady. "I tried to pass it off and figure, 'This is going to go away. I can't be having all these symptoms.' I thought I was going nuts."

In early 1984, the Stecks found a mimeographed flyer in their local newspaper urging residents to attend a town meeting scheduled to discuss problems related to Yukon's "acid plant." Neither Gary nor Diana knew what the acid plant was, what it produced, or even precisely where it stood. But Gary decided to attend the meeting since several of the town's oldest and most prominent families backed it.

Gary and Diana had heard about the cattle that died back in 1972 while grazing next to the acid plant. The ranchers sued the state, alleging its regulatory agency had allowed the acid plant to contaminate the groundwater. In more recent years, Yukon residents objected to the smells wafting over the facility's walls and the yellow cloud that sometimes crowned the site, burning the eyes of passersby and filling their mouths with a bitter, metallic taste. Even more residents decried the extraordinary flow of traffic. On some days, as many as 40 trucks wove through Yukon's narrow streets to negotiate entry at the main gate of the plant.

Gary returned home from the meeting confused and dispirited. The evening had proved chaotic, everybody shouting and talking at once. In the end, nobody could offer solid information regarding the acid plant's operations or a plan for making it safer. Gary told Diana that he couldn't even tell if the acid factory presented a problem. "If the folks at the meeting were doing something," he told Diana, "I'd help. But nobody's doing anything." And the Stecks had more urgent problems at home. In late 1984, Jeremy contracted nonviral hepatitis. A few months later, Diana was tentatively diagnosed with lupus. She went on leave from her nursing job at the hospital. "We completely put the acid plant out of our minds," said Diana.

In the winter of 1985, almost a year after Gary attended the town meeting, Diana read in the local newspaper that the Department of Environmental Resources planned to close down a hazardous waste treatment and disposal site located in Yukon. The article listed some of the chemicals contained at the waste site, describing them as "pickle-liquor." As the daughter of a steelworker, Diana knew the term. Pickle-liquor referred to a bath of sulfuric, hydrochloric, and nitric acids used in the mills to cleanse impurities from the steel.

When Gary came home from work that night, he slipped into his chair, picked up the newspaper, and read the same article. "What is this?" he asked. "Is this that acid plant?"

Diana didn't know. "I want to find this place and see what it looks like," she told Gary. They drove around town until they located the site. The acid plant and the hazardous waste dump were one and the same.

"I will never forget looking at its walls for the first time," recalled Diana. "It was leaking all this stuff out the side, and it was running down across the road. It just looked watery, but it had a horrible smell. My nose and my mouth burned."

Then an idea occurred to Diana that would change her life.

"All of a sudden, something clicked. I said, 'It's this place. This is why we've been sick.' "

The next day, Diana called another nurse at work and asked her to check out a toxicology text from the hospital library and bring it to her home. Diana stayed up the entire night reading about the chemicals listed in the newspaper article. She found out that pickle-liquor contained chromium, cadmium, lead, arsenic, and other elements that could cause cancer, chronic bronchitis, skin lesions, gastrointestinal problems, nerve damage, and additional illnesses.

"All the crazy symptoms and the blood tests that we all had that didn't make any sense now suddenly made sense," said Diana. "Every one of our symptoms was listed. I was convinced."

The possibility of a connection between the waste site and her family's illness terrified Diana, but she took some comfort in the state's order to close the dump site.

" 'That's it,' I thought. 'They're going to shut them down. Jeremy's going to get better. We're all going to get better and we'll live happily ever after.' "

But the image of the leaking hazardous wastes haunted her. Closing the dump made no difference if the toxic chemicals stored inside would continue to harm her family's health. Somebody had to clean up the contamination.

The Department of Environmental Resources had ordered the owners, Mill Services Inc., to devise a closure and cleanup plan. To Diana, this idea sounded crazy since the company had not safely contained the wastes in the first place. Diana decided to inform the government officials about her family's health problems, matching their symptoms with the chemicals still accumulating at the dump. Once the government understood the public health threat, she reasoned, the appropriate officials would move swiftly to erase the contamination.

In the hills above Yukon, a lake was created from liquid hazardous wastes containing chromium, cadmium, lead, arsenic, mercury, and dozens of other toxic chemicals.

Diana sat at her kitchen table to thumb through the Yough County telephone directory. She started with the local elected officials, telephoning and writing letters to the county commissioners. They responded cooly, informing Diana that responsibility for the matter lay beyond their realm. Diana worked her way through the phonebook, calling her state assemblymen and state senators, the Department of Health Services, the Department of Environmental Resources, her United States congressman and senators.

At each office, the officials seemed to be reading from the same script.

"Everybody told me, 'Gee, Mrs. Steck, I really understand how you feel. We'll look into it. We'll get back to you.' "

Nobody called her back. Diana would wait a week, and then call again.

"You told me you were going to get back to me!" she barked into the receiver, surprised by her own boldness. "What did you find out?"

The politicians, their aides, and the staff people at the state environmental regulatory agencies commiserated with Diana, but nobody promised to do what she wanted: clean up the dump and move its wastes elsewhere. Several health department officials pointed out that Diana had no proof that the dump bore any relation to her family's illness. They informed her that people get sick all the time, especially children. Not every family tragedy could be blamed on the toxic waste dump next door. Besides, the government knew about Yukon's problems. Qualified experts would review the case.

Diana wasn't satisfied. She wanted an established government agency—staffed with scientists, lawyers, policy experts, whomever the task required—to assure her she could stop worrying about her own health and her son's condition. She needed an ally, somebody—unlike herself—with power.

For six months, Diana navigated the labyrinthine channels of local, state, and federal government. The Steck household kitchen took on the appearance of an amateur lobbyist's office. In the morning, Gary would leave for work while Diana pounded out another letter on the kitchen table typewriter. When he returned in the evening, Diana would be clinging to the telephone, tracking down the name of another government staffer holed up somewhere in the bureaucracy. The Stecks' telephone bill climbed to $500, then $600 each month. Gary couldn't decide whether he resented or admired his wife's new obsession.

"One day, after six months of beating my head against the wall and feeling totally alone and helpless," recalled Diana, "somebody at the EPA gave me a phone number. He said, 'Honey, I can't help you. But this person can. Call her.' "

The phone number belonged to Lois Gibbs at the national office of Citizen's Clearinghouse for Hazardous Wastes in Arlington, Virginia.

Diana had never heard of Lois Gibbs or the Citizen's Clearinghouse for Hazardous Wastes. But she did remember watching a television documentary about Love Canal, the suburb in upstate New York contaminated by its neighbor, the Hooker Chemical Company. Lois Gibbs had lived in Love Canal; she was a housewife with a sick child who threw herself into the fight to force the government to evacuate her neighborhood and clean up the chemical wastes. Since Love Canal, Gibbs had risen to prominence as the founder and director of the national anti-toxics group, Citizen's Clearinghouse for Hazardous Wastes.

"In the documentary," recalled Diana, "I remembered the swing set and the house with the white picket fence all boarded up. Jeremy was an infant then, and I watched it while I was sitting in the rocking chair and feeding him his bottle, thinking, 'Oh, my God, those poor people.' "

Diana called Citizen's Clearinghouse for Hazardous Wastes, breathlessly insisting that she was "supposed to talk to Lois Gibbs." The person on the other end of the line cooly informed Diana that Lois Gibbs was busy.

"I have to talk to her," demanded Diana. "I absolutely have to talk to her. Put me through to her." Six months of tracking down politicians and badgering government bureaucrats had invigorated Diana's phone manner. She could prove particularly obstinate when she felt that she was being dismissed.

Lois Gibbs picked up the line. She and Diana talked for an hour.

The most terrible and unforgettable part of the conversation revolved around their children.

Lois explained to Diana how she had moved into Love Canal with her husband and bought the American Dream. The Gibbs owned a small ranch house, two cars, and a color television with HBO. They also had a one-year-old son named Michael. Soon after moving into the new subdivision, Michael took ill. At first, he suffered from a flurry of skin rashes, ear infections, coughs, allergies—normal childhood problems—but for Michael these illnesses were much more frequent and severe than for other kids. After four years in Love Canal, Michael developed severe asthma, epilepsy, liver ailments, and a urinary tract disorder that took two operations to correct.

One day Lois' second child, Melissa, suddenly developed saucer-sized bruises all over her body. The blood vessels beneath her skin burst. Her mouth and nose bled profusely. At first, the doctors thought she had leukemia. During a bone marrow test, Melissa had to forego anesthetic since her blood-clotting ability was so limited that she might have bled to death. When the doctor pulled out a 12-inch needle to insert into the middle of the child's bone, Melissa begged her mother to take her home. She insisted that she hadn't done anything wrong, she didn't understand why the doctors wanted to hurt her.

By that time, Lois Gibbs was convinced that the chemicals surrounding her home were the cause of her children's illnesses.

Diana listened to Lois in rapt attention. "It was the first phone call in six months that anyone knew what I was talking about," she said. "I didn't feel alone."

Diana confided in Lois about Jeremy's urinary tract infections, her own joint pains, fatigue, the numbness in her limbs. Lois listened attentively. She told Diana how she, too, had talked to officials at every level of government, trying to warn them about the health risks posed by her town's chemical contamination. They hadn't taken her seriously either. So Lois Gibbs tried something different. She knocked on her neighbors' doors. She talked to anybody who would listen about her family's health problems and asked perfect strangers if their own families were ill. She slowly uncovered terrible stories about birth defects, miscarriages, and stillborn children.

Diana Steck stared at her children's toys spread across the floor. She sat at the kitchen table, listened to Lois Gibbs, and gripped the telephone receiver in her fist.

- - -

Diana Steck and Lois Gibbs were right to worry about their kids. Toxic chemicals pose a vast range of dangers to children's health, often making them the first victims of a community's contamination. For a variety of readily apparent reasons, children are more susceptible to the effects of chemical exposures than adults. At the heart of the problem stands one obvious truth: Children are not simply shrunken adults. Their behavior, diet, developing physiology, and growth needs all contribute to children's unique susceptibility to toxic harm.

The mere fact of children's short stature increases their jeopardy. Kids live closer to the ground than adults, maximizing their contact with toxic substances that collect indoors. They may routinely encounter small quantities of household chemicals spilled in the kitchen or bathroom, solvents dragged into the house from the garage on shoes and work clothes, or airborne garden pesticides wafting through an open window. The family toddler, who spends his days noodling across the living room floor, may stir up and breathe chemical residues as he grinds his body into a toxic carpet that has served as gravity's last dropcloth.

The world outdoors also appears considerably more poisonous when viewed from the perspective of childhood. In Idaho's Silver Valley, two hundred children developed acute lead poisoning, many requiring painful hospital treatments to remove the heavy metal from their blood—all from playing outside in the dirt. The circuitous route of contamination shocked parents and public health officials. Nearby industrial smelters spewed lead into the air; the kids rolled around in the dirt where the lead had settled: Before they got a chance to wash their hands, they stuck their fingers in their mouths.

In other communities, children's innate curiosity has drawn them to the center of toxic hazards. In Lowell, Massachusetts, the town's contaminated dump attracted scant notice from the adults passing by each day on their way to work. But kids walking to school often stopped to spark matches upon the chemically saturated soil and light the ground on fire as a game. This kind of child's play served as one explanation for why Lowell's children suffered from high rates of colds, sore throats, and ear infections—an indication of their weakened immune systems. In Columbia, Mississippi, kids flocked to an unguarded chemical dumping ground as though the "No Trespassing" posters tacked up along the perimeter were neon signs advertising free rides on an amusement park Ferris wheel. "There's gooey stuff and you can squish and play with it," enthused a six-year-old boy, referring to the dump's mud pools soaked with leukemia-causing benzene.

Beyond the paradoxical attractions of toxic dumps, kids come into greater contact with dangerous chemicals simply because they eat more than adults in proportion to their body weight—including more pesticide residues from fresh vegetables, fruits, and juices. The EPA classifies as possible carcinogens over 65 percent of the 560 million pounds of herbicides and fungicides sprayed annually on U.S. crops. The average child consumes four times the amount of these suspect chemicals than an adult.

Kids' immature digestive tracts make matters worse. Adults will absorb only 10 percent of the lead they accidentally ingest; children absorb 50 percent. While adults eliminate half of the lead introduced into their bodies within one month, children must wait ten months to achieve the same rate of detoxification. And kids prove far more likely to encounter this extremely dangerous element on a daily basis. Not only do children play in the lead-contaminated dirt of our cities, but nearly 900,000 units of public housing and 57 million homes have been covered with lead-based paints whose flecks and chips too often end up in the mouths of young children. Federal investigators also estimate that hundreds of thousands of kids drink lead-contaminated water at their schools.

Since even low levels of lead can damage hearing and interfere with a child's developing nervous system, it is not surprising that kids exposed to lead at an early age prove seven times more likely to exhibit learning disabilities. Lead poisoning in children has also been linked to convulsions, brain degeneration, and death. Dr. William Roper, director of the federal Centers for Disease Control, asserts that "lead poisoning is the number one environmental problem facing America's children." But other chemicals may offer comparable dangers. "We've been studying lead for a long time," pointed out Dr. Herbert Needleman, the nation's foremost researcher on childhood lead exposures. "We're twenty years behind on the study of pesticides. . . . There is no question that pesticides impair children's brain functions as insidiously as lead. I will tell you without fear of contradiction that exposing children to excessive levels of pesticides is impairing their health, eroding their mental abilities, and shortening their lives."

Ironically, parents may provide the most direct route for dangerous chemical exposures. When the EPA examined a group of young mothers for contaminants in their breast milk, they discovered that 99 percent of the women carried traces of the pesticide DDT and cancer-causing PCBs. Another study found 190 chemicals in breast milk, including leukemia-causing benzene and the carcinogenic sedative, carbon tetrachloride. Despite its critical importance to nursing babies, the purity of human breast milk on average fails to meet the standards set by the Food and Drug Administration for commercial cow's milk. At a time in life when an infant's developing enzyme system and digestive tract allow for the highest possible levels of absorption, breast-fed babies take in six times more chemicals than the EPA considers safe for adults.

Even prior to breast feeding, many children contend with a toxic environment within their mother's womb. Ninety percent of the U.S. population carries in its bloodstream and fatty tissues traces of an electrical insulator, widely used in industry for over 40 years, called PCBs. During pregnancy, PCBs cross the placenta to reach the developing fetus. Babies born to mothers with high levels of these chemicals evidence sluggish muscle movement and decreased reflexes, indicating a more slowly functioning nervous system.

For years, scientists ignored the father's role in passing along a dangerous chemical legacy. "You don't have to be Sigmund Freud to figure out there are cultural factors to say why we have paid so much attention to the female and so little to the male," noted Dr. Devra Lee Davis, an expert on reproductive toxicology at the National Academy of Sciences. But current research shows that men pose a far greater threat to their offspring, both at conception and throughout childhood, than most scientists had previously suspected.

Prior to birth, the culprit is the sperm. Researchers once thought of the sperm as a simple transport vehicle carrying male DNA to the egg. But scientists now believe that the genes in sperm can become mutated by toxins. These mutations, delivered by the sperm, become part of the genetic makeup of a developing fetus, damaging the embryo at its earliest stages. And recent investigation also shows that the chemical barrier separating the blood from the testicles is very permeable, enabling toxic chemicals to easily intermingle with developing sperm, some of which is finally delivered to the woman's egg with the ejaculate.

All this underscores the fact that the effects of toxic exposures at the father's workplace have been greatly underestimated. The *Journal of the National Cancer Institute* recently reported that parental contact with workplace solvents increases the likelihood of children under ten contracting leukemia. Welders who inhale toxic metal fumes can produce damaged sperm up to three weeks after the exposure ends. Toxic smoke inhaled by firemen elevates the risk of fathering children with heart defects. Garage mechanics, automobile body shop employees, and other transportation industry workers, who consistently labor amid hydrocarbons, solvents, metals, oils, paints, and other cancer-causing petroleum products, bear a four-to-eightfold heightened risk of passing along genetic material that will give their children kidney tumors.

Chemicals employed in the workplace can also travel home on either parents' skin, clothing, and even in their breath. Like the lingering odor of alcohol long after the last drink, solvents inhaled by parents at work continue to be excreted in their breath for hours after returning home. The affectionate kiss between parent and baby at the end of the workday may also offer the child a regular dose of benzene from petroleum products handled on the job.

It may seem strange that children should suffer risks from low-level chemical exposures while parents largely escape their effects. But children's rapid growth process makes their bodies far more vulnerable to the influence of toxic materials. In the fetus, cells must divide quickly, opening them up to genetic damage. Among newborns, whose bones and brain develop more rapidly than at any other time in life, this heightened sensitivity continues.

Finally, today's children bear one more burden: They will live longer than their parents in an increasingly toxic world. Given the vast number of synthetic chemicals introduced into the environment since World War II, the children of the late twentieth century will absorb a lifetime of toxics at levels far greater than those encountered by their parents or grandparents. From conception to the grave, it remains impossible to accurately assess the full weight of the chemical load carried by the children of Diana Steck, Lois Gibbs, and countless other mothers throughout the country.

■   ■   ■

"You're going to have to talk with your neighbors," Lois Gibbs told Diana Steck.

Throughout the six months Diana had been searching for an ally, she hadn't once spoken with other Yukon residents. It had never occurred to her.

"I can't do that," objected Diana. "I can't talk in public."

One day over the telephone, Lois walked Diana through her fears. Lois confessed that she had also been afraid to strike up a conversation with strangers, particularly regarding a matter as intimate as her family's health. So before she started knocking on doors in Love Canal, she composed a short speech and forced herself to rehearse the lines. It had been an excruciating exercise and she felt ridiculous—but it worked. Lois owned a dog named Fearless, and several times each day she would shake his paw, gaze into his panting face, and repeat her speech over and over and over: "Hi, my name is Lois Gibbs and I live at 535 101st Street, and I'm really concerned about my kids." When Lois' husband returned home from work, she would grab his hand, shake it, and tell him the same thing. "I know who you are," he'd answer, "I sleep with you every night." After several weeks, she worked up the confidence to hit the streets.

Diana could do the same. She had the one best reason to try—her kids.

"There are only two sources of power in this country," Lois lectured Diana. "One is money. You and I don't have any. The other is people. This really is a democracy, a country of the people, by the people. If the people choose to take it."

The next day, Diana returned to her kitchen table typewriter to hammer out a makeshift petition. It enumerated the dump's potential health hazards and urged state officials to launch an investigation. Diana didn't fret over the document's wording. As Lois advised, the petition would serve mainly as an excuse to wander around town and talk to the neighbors. But by chance, Diana's neighbors came knocking first.

On Diana's porch stood Angie Babitch, one of the organizers of the chaotic public meeting attended by Gary more than a year before. Angie rustled a sheaf of papers in her hands and explained to Diana that she and the Bolk family, who had sued the Department of Environmental Resources years before over the dump's groundwater contamination, now planned to go door-to-door to survey Yukon residents about their health problems. After a decade spent shrugging off complaints, Yough County's health department now urged the Bolks and Babitches to launch their own investigation to determine if Yukon suffered from high rates of disease. If the officials received a convincing list of symptoms, Angie told Diana, they'd do something about the dump.

Diana didn't have much faith in the local health department, but she knew immediately that she would work with the Bolks and Babitches. Not talking to her neighbors had been a colossal mistake. Diana joined Angie Babitch, and they soon hit every house in town. Their informal survey turned up complaints about skin rashes, asthma, bronchitis, and urinary tract infections that seemed to resist treatment with antibiotics. Diana and her friends theorized about wind-borne fumes or particles sweeping through town from the dump. They thought the chemicals might be working their way through the soil to contaminate the ground beneath people's homes.

Three weeks after their door-knocking began, Diana piled into a car with Angie Babitch and her husband, Nick, and Melbry Bolk, and drove for five hours to Hawking Hills, Ohio, to attend a leadership training conference sponsored by VOICE, a statewide group of grassroots environmental organizations. Gary had pushed Diana to attend the conference. "If you're really going to do this," he told her, "maybe you can get some help." Whatever it took, he wanted his wife to finally get this problem behind them.

On the first morning of the conference, Diana nervously entered the hall to encounter several dozen people from Ohio and Pennsylvania sitting quietly in their folding chairs, or drinking coffee and chatting—but otherwise looking as uncomfortable and uncertain about their presence at the meeting as Diana now felt. The women wore plain cotton dresses or their best slacks and blouses, and the men dressed in blue jeans, flannel shirts, heavy work boots, and the ubiquitous baseball and feed caps that had become the trademark of male America. Their voices rang with the twangy resonance of the rural Northeast, a soothing and familiar sound. Diana recognized the roomful of strangers as small-town housewives and mothers, farm families and working people; they were like her and Gary, her parents, the people she had grown up with in Jennette who had never before been involved in protests or demonstrations or politics of any kind. For the first time, it hit Diana that Yukon's troubles were not unique.

As the conference began, Diana felt too bashful to speak. Nick Babitch stood up and introduced the Yukon contingent.

"Diana's a transplant," he explained to the crowd. "She's not from Yukon all her life."

Nick's comment upset Diana: She would always be an outsider, he seemed to be saying. But as he spoke, Diana suddenly recognized that her peculiar status wasn't necessarily a handicap in the struggle against the dump. "Because I wasn't born in Yukon," she explained, "I was more apt to look for a reason for what was causing the problem. When you don't have your roots in a town, you're more likely to want to make changes. You don't care too much about going against tradition."

By the conference's end, Diana felt certain that she had to involve even more people in the fight. She returned to Yukon, spoke once again with the families she had met canvasing door-to-door for the health survey, and recruited 12 members to form CRY, Concerned Residents of Yough County. From the beginning, CRY's tactics improved upon the previous decade's stumbling efforts. CRY quickly adopted a single, clearly stated goal: Shut down the dump and clean up the mess. Recognizing how Yukon's old, established families had dominated previous attempts to prod local government into action, Diana insisted that membership be opened to the entire community. She invited people she barely knew to attend meetings at her house. Word spread throughout town and every gathering included new faces.

But CRY had to invent itself as it went along. No set of rules, no guidebook nor secret formula existed to compel the government to conform to its citizens' desires. Diana and CRY made mistakes.

At first, the organization veered towards "the technical track," scurrying around the state to recruit sympathetic experts to prove the dump had harmed Yukon's residents. Diana found hydrologists to chart the flow of the leaking waste site into Yukon's groundwater. She contacted geologists to impugn the dump's stability. She talked with toxicologists and epidemiologists who could attempt to link the symptoms turned up by the health survey with the toxic contents of the dump.

Citizen's Clearinghouse for Hazardous Wastes maintained a roster of sympathetic experts, as well as a "hit list" of uncooperative technical people, whose findings consistently sided with government and industry. But Lois warned Diana off the technical track, pointing out that for every expert the community could hire, the opposition would present three others to refute his or her findings. In the end, Yukon would wind up with "dueling experts," as competing cadres of white-smocked scientists waved their flowcharts and Ph.D.s at each other, providing government officials with an excuse to throw up their hands and do nothing.

"When somebody asks me a technical question," said Lois, "I tell them, 'My specialty is in not knowing that.' The battle is political, not scientific."

But Lois didn't expect Diana to take her word. The technical track invariably seduced communities at the beginning of their fight with promises of a quick, painless resolution. Instead of arguing with Diana, Lois suggested that she call Penny Newman in Glen Avon, California, who had been battling the Stringfellow toxic dump since 1979. Over the phone, Penny explained how her own community devoted years to the experts, largely to no avail. Steps to clean up the leaking dumpsite above their town only commenced once the community group agreed to target the decision makers—pressuring, criticizing, embarrassing, and otherwise harassing the politicians who held sway over the state and federal regulatory agencies. Diana hung up the telephone, convinced that Penny was right. Her own limited experience with the technical experts had cannibalized her time— and ultimately, proved fruitless. Don't trust the experts, Penny had warned her; depend on your family, friends, and neighbors. Trust the people who know and care about you.

Diana liked Penny nearly as much as Lois. She had never met the woman, but she felt in Penny the force of her intense good humor and seriousness. Other telephone friendships blossomed. Now whenever Diana called Citizen's Clearinghouse with a question, Lois referred her to somebody else in another part of the country—almost always a mother and housewife much like Diana, or Penny, or Lois herself—who had already dealt with the same problem. Diana spoke with women all over the United States whose advice and experience guided her decisions. Although she seldom met these women in person, she felt their presence. She was not alone.

With mounting enthusiasm, CRY clamored down the political path. The organization held highly publicized candlelight vigils, noisy public protests, even a Christmas parade. Ten CRY members blocked the entry to the dump for three and one-half hours by lying down in front of incoming trucks while another 40 people cheered them on. Newspaper and television reporters showed up to cover the events. Conservative, rural, western Pennsylvania seemed an unlikely hotbed of political protest; every small act of rebellion earned headlines.

Throughout 1986, CRY breathed new energy into Yukon. Street captains took charge of every block, spreading the word about the next public meeting or demonstration; they assigned and then followed up on the letters members had to write by Monday, the telephone calls to be completed after work. Since CRY didn't have any money, personal contact proved cheaper—and faster—than the mail. As important, it kept people talking to each other.

Diana volunteered her house as the CRY office. People filtered through its doors day and night, eroding Yukon's overgrown respect for privacy. Diana devoted 12 hours a day to the fight, leaving Gary at home evenings and weekends to cook and take care of the kids. During the day, she transported her two year old, Jessica, to all the public hearings, strategy sessions, conferences, and protests. She discovered that she could cut short the time wasted in any public official's waiting room by handing her toddler a large, sticky chocolate bar and allowing her to rub her hands wherever impulse dictated. Yet as the kids grew older, they often rebelled against the incessant meetings and demonstrations. "You love the dump more than us," both kids would sometimes complain. "Why can't we have a normal life?"

Since CRY had formed, Diana had changed so much that even Gary sometimes couldn't recognize her. His wife's stubborn streak had blossomed from a familiar, quiet obstinancy into a brazen determination to stand and fight for what she believed. "When your mother died, she gave you a gift," Gary told Diana with grudging admiration. "She gave you her mouth." Other times, he bristled and complained more fervently about the constant activity around the house, wondering why the Steck family had to take on the entire world for the benefit of Yukon. Diana worried about the dump fight eroding her relationship with Gary. She had heard from other women around the country how their own anti-toxics activism had proved corrosive, even fatal, to their marriages. And Diana knew that she needed Gary's moral support. Finally, she approached her husband strategically: She replaced the newspapers and magazines stacked up in the bathroom as reading matter with technical reports about the dump. Gary had majored in biology at college; he could interpret the reports and they alarmed him. In this manner, Diana gradually seduced her husband into joining the fight alongside her.

Other people had even greater problems with the new Diana.

"I don't know what happened to you," observed an old college friend, "but somewhere between nursing school and now, you flipped!"

"Is it normal for somebody to get out there and try to stop trucks on the road?" demanded Diana's best friend from high school.

"If somebody was poisoning your children," Diana shot back, "you'd do the same thing."

Only a year before, Diana would have felt crushed by the criticism of her old friends. Now she had begun the painful, if inexorable withdrawal from their regard. "I knew they were the ones who were screwed up," she said. "All this happened for a reason. Otherwise we'd still be out there, just stupidly working and making money, oblivious to the world around us. We wouldn't be the people we are today. We wouldn't be as complete."

As the fight against the dump picked up momentum, Diana learned about other communities involved in toxic waste issues. In East Liverpool, Ohio, the people were fighting an incinerator. In Ashland, Kentucky, another anti-toxics group resembling CRY had formed. In other parts of Pennsylvania, Michigan, Indiana, New Jersey, New York, Virginia, West Virginia, and Illinois, people were fighting—and sometimes winning—the same battles that were consuming her life in Yukon.

Almost always, women led the fight in these towns—usually mothers with kids at home. To Diana, it made perfect sense. Women, not men, cared for sick kids. Women cooked the meals, drew the water for their children's baths, watched their kids play outside. The chemical threat that poisoned food and water, and turned local playgrounds into toxic heaps, struck at the heart of the woman's realm.

Over time, Diana noticed pointed similarities between herself and the leaders of other contaminated communities. Typically, the women who sparked action viewed themselves as outsiders. They weren't necessarily troublemakers or rebels or malcontents—but somehow they didn't quite fit in. Like Diana, many came from someplace else; in Nick Babitch's term, they would always be "transplants." Others distinguished themselves with a college degree or professional job—such as Diana's position as a nurse. Their families usually contained at least one sick person, sometimes the woman leader herself. Previous organizational or political experience, if any, extended no further than membership in church committees or the PTA.

The women leaders of small-town toxic America did not draw inspiration from the theories of deep ecology, Green Party socialism, enviro-anarchism, or eco-feminism; they were Democrats and Republicans, or just as often, they belonged to the great nonvoting, apolitical majority who presumed the government to stand far beyond their reach and influence. They relied on their experience as housewives and mothers to inform their view of the world and to shape their tactics and strategy once the government they took for granted failed to protect them and their families.

Diana also noticed the profound differences between the anti-toxics community groups led by women and the institutions that she had dealt with all her life—the schools, hospitals, and governmental agencies, inevitably dominated by men. The new world of grassroots organizations in which she now immersed herself drew sustenance directly from the intense relationships forged between women, making their work all so ... *personal.* Diana didn't regard Lois Gibbs merely as an ally or the leader of a political movement; she was a friend. She knew the details of Lois' past, her family's history; sometimes Lois felt like her older sister. In turn, Lois acted less like a model for Diana to emulate than an intimate and equal from whom she could draw solace and strength because of their similarities.

Diana often talked about "the caringness" of the women she encountered in groups like CRY. This quality connoted more than

Diana Steck with members of CRY.

sensitivity, sympathy, or concern. It spoke to an aggressive inclusiveness, the willingness of one person to take on another's pain. When the women in other contaminated communities spoke, Diana heard her own story. Every time they emerged victorious from another skirmish with the local health department, EPA, or the polluters, her own self-respect and confidence flourished. The rampant empathy of the anti-toxics movement enabled members to face down the intimidation and casual disregard of doctors, public health administrators, government regulators, and politicians. The experts and officials could lean back in their swivel chairs amid the comfort and safety of their fancy offices and argue the rules and regulations all day; but the women kept asking: "How does any of this protect my family?" The women could endure the repeated humiliation of being written off as "hysterical housewives" by Ph.D. biochemists from the most prestigious universities, millionaire captains of industry, the highest-ranking and most powerful politicians—the people in American society who mattered most—all because what mattered most to these women was their children. They could not believe that the arrogance and posturing of the privileged authorities protected their kids in the least. While men in the anti-toxics movement might identify fleetingly with the officials, the women never had any illusions that in a different life, they would be the ones wielding power and making up the rules. The hard lessons of class and gender—not in the abstracted realm of theory, but in the concrete facts of their daily lives—informed these women that they were not the primary actors in the unfolding American drama, but the people most often acted upon.

As newspaper accounts spread about Yukon's fight against the toxic dump, women in rural Pennsylvania, and parts of Ohio and West Virginia, read about Diana Steck and called her at home for help. Over the telephone, they spoke about their fears for their children, their frustration over government inaction, their uncertainty about what to do next. Some broke down into tears of rage and sorrow. Diana offered advice, traveling at her own expense to help organize several communities. Diana's view of the world expanded.

"At first our attitude was, 'I just care about Yukon,'" admitted Diana. "Who cares about what's going on in some other part of Pennsylvania or Ohio or anyplace else?" But excursions to other small towns that could have passed for Yukon's twin sisters forced Diana to care. In every community, she found young mothers, much like herself, who needed her help as badly as she had needed Lois and Penny. Alone, they could do nothing. Working together, Diana sometimes thought they were invincible.

But Diana couldn't dally on the road for long. Back in Yukon, CRY had its own problems. Foremost, Diana had failed to anticipate the influence of her own example. "I was just plowing ahead. I could give 120 percent, and I expected the same of everybody else. I was so driven by the need for quick action that I couldn't wait until everybody caught up with me." Every time CRY considered a new tactic or tried to clarify an issue, members turned to Diana for guidance. Once ignored on the streets, shunned in her own church, Diana now set the town standard for dedication to the cause. In 95 percent of CRY's debates, her point of view prevailed. In the context of community organizing, this kind of influence did not constitute success; in truth, it meant that Diana had overstepped her role as organizer, prodding people into positions they probably weren't ready to adopt.

For example, when CRY held a meeting to formally set its goals, Diana persuasively outlined what she considered wisest.

"Now what do you think?" she asked.

Of course, most people thought what she thought. Who were they to disagree? Nobody in CRY had ever battled big business or tackled the government before. They didn't know how to publicly disagree or work out compromises on strategy and tactics. They kept their doubts to themselves. And few people could tell that Diana wasn't nearly as certain about her opinions as she often seemed. In fact, her conviction usually served as a means to battle the immense sense of helplessness that sometimes overcame her. She hated being in the middle of a situation that seemed out of control.

Finally, while Diana could inspire people, she also intimidated them. And her zeal too often smothered the emergence of other, critically important town leaders.

"I thought I was being democratic," confessed Diana. "It was more a dictatorship by volunteerism."

Six months later, CRY reexamined its goals. At this point, Diana believed Yukon should push for money from the government, the dump's owner—whomever—to relocate residents to a new town, as groups had managed in Love Canal, Times Beach, and a handful of other communities. But the membership didn't share this goal. In some sense, Diana was still an outsider; the rest of the group hoped to reside in Yukon for the rest of their lives. This time, Diana fought back the temptation to out-argue her colleagues, and she wisely dropped the issue.

In truth, Diana was a natural organizer, and most of the time her instincts served her well. Since she didn't know the rules of political organizing, she broke them without regret. When a tactic failed, she proposed something new—and if it worked, she kept it up until something else occurred to her. It didn't matter what anybody called it. Early on, for instance, CRY dispensed with an old standard of community organization, the committee structure.

"When you say committee in Yukon," explained Diana, "everybody heads for the hills. You can't use that word." Instead, CRY leaders emphasized the task that needed to be accomplished. "We don't say, 'Let's have a meeting of the fund-raising committee.' We'll say, 'How about three people getting together and doing a bake sale?' Three people will volunteer, they'll do a good job, and then continue to do all the bake sales. Sure, we have committees. We just can't call them that."

Informality mixed well with CRY's sharply focused goals, lending the organization both openness and stability. As a result, CRY swelled with new members. At one public meeting, five hundred people showed up. Diana felt ready to take on the state directly.

In August 1986, the Department of Environmental Resources finally agreed to hold a town meeting in Yukon to review the closure plan for the dump. The state wanted to install a clay cap over open ponds, while constructing another lagoon to store additional wastes. CRY insisted that the entire dump be closed and cleaned up. Diana called Lois to discuss tactics.

"You have to take charge of that meeting," warned Lois.

Diana and Lois both knew that the officials who convened public meetings invariably controlled them. They could turn the hearing process into a safety valve for community pressure, allowing citizens to parade up to the speaker's podium to vent their anger, fear, and frustration—and waste their one opportunity to force the officials into a public commitment. The bureaucrats' capacity for absorbing abuse could be wonderful to behold. They just sat there, and took it, and didn't promise anything. The routine gave the impression of serious and concerned public servants listening carefully to the people. In reality, it proved another triumph of the flak catchers. After the people had their say, the bureaucrats returned to their offices and did exactly what they had planned from the start.

A disciplined community might turn the situation around. Lois told Diana about the public meeting in Love Canal when the Homeowner's Association confronted New York governor Hugh Carey, three months prior to his bid for reelection. The community group knew from the problems and failures of past meetings that when numerous people spoke publicly, one person's worries merely drowned out another's. The officials nodded in apparent sympathy and then moved on to the next speaker. And so the Love Canal group agreed this time to focus their demands with a single speaker. Nobody else would raise a hand to speak until the governor answered the first question. At the beginning of the meeting, Lois marched to the front of the auditorium, followed by a brigade of three-, four-, and five-year-old children who formed a ring around the podium. "Are you going to allow these children to remain in Love Canal," asked Lois, "*and die?*"

The governor ignored this hysterical housewife, craned his neck around the obstruction of children gathered in front of him, and scanned the audience for a more rational citizen with a polite and reasonable question. Nobody raised a hand. Reporters' cameras flashed. "Sir, sir?" demanded Lois. "Could you answer the question?" After an excruciating silence, Carey responded. Speaking spontaneously—much to the horror of his aides—he pledged to evacuate the 239 families who lived closest to Love Canal's most contaminated area.

When 350 Yukon residents piled into Slovenian Hall to meet with state officials, the community proved equally well-organized. The state had demanded a list of the people who wished to testify. Diana headed the list. She walked across the stage where the officials sat at a long conference table. Somebody from the Department of Environmental Resources handed her the microphone. Diana spoke briefly, explaining that the people of Yukon gathered in row after row of folding chairs below the stage all vehemently opposed the closure plan. When she finished speaking, the man from the DER reached out his hand for Diana to return the microphone.

"I wouldn't give it back," remembered Diana. "We'd planned that I would not relinquish the microphone and I would take over the meeting. I called the next person."

The hearing chairman pounded his gavel. Diana ignored him. Speakers proceeded to the front of the auditorium. The emotional testimony of local residents mixed with the technical experts who outlined the dump's structural flaws and health threats. When somebody in the audience wanted to make an additional point, Diana called him or her up to the stage. "The evening just flowed," recalled Diana. A group of Yukon kids performed a Humpty Dumpty skit, moralizing that once damage is done, there's often no way to fix it. The skit concluded with the symbolic cracking of an egg on the floor. While the kids swabbed it up, Diana commented dryly: "See, our children aren't like the dump's owners. They clean up their mess."

Some months earlier, CRY members had learned from a local reporter that a state official had labeled them "country bumpkins." So when two young girls climbed onto the stage wearing white lab coats with "Country Bumpkin" signs pinned to their backs, the room exploded into laughter and applause. The girls conducted a mock science experiment, placing a lid over a flowerpot filled with dirt. But the flowerpot was riddled with holes at the bottom. Dirt dribbled all over the floor, a symbol of Yukon's attitude about "putting a lid over the leaky bucket" at the Mill Service dump.

Dirt fell everywhere. When the girls sprinkled some soil across the officials' conference desk, the chairman whacked his gavel and informed the community that the meeting was over. The committee was going home.

Diana hadn't considered this possibility, but she had no intention of surrendering. She stood up, grabbed the microphone, and faced the audience.

"Gentlemen," she commanded, "bar the doors."

The burly ex-miners and farm hands scattered throughout the audience didn't have to rise from their seats because several elderly women had already crossed the auditorium floor to lock the doors. The state officials stared out unbelievingly at the crowd. Their implacable faces glared back anger and determination. The officials sat frozen at the conference table elevated high above the audience, suddenly on display. Nobody moved; they had no intention of leaving. They had totally lost control.

Diana called the next speaker and the meeting proceeded. Yukon's citizens hectored their captives about public responsibility, governmental openness—in a word, *democracy*. At the hearing's conclusion, Diana read CRY's list of demands. CRY wanted to be treated as a third party in cleanup negotiations between the company and the government. The community organization required copies of all correspondence and documents, free of charge. It insisted on a new public hearing to review the revised closure plans. An audience member handed each state representative a pair of flash cards reading either YES or NO. After each demand, Diana asked the committee members to hold up the card indicating their response. The committee acceded to every demand.

They had won an important, invigorating battle—but the war would continue.

"Who Do You Wave Your Sword At?"
—McFarland, Winter 1988

McFarland, California. Guillermo and Tracy Ramirez with their daughter, Angela, who developed cancer when she was two. After chemotherapy and radiation, the cancer was in remission.

McFarland's parents feared for the lives of their children.

The deaths of Mario Bravo, Mayra Sanchez, Tresa Buentello, and Franky Gonzales caused many mothers to scrutinize their own families each day for the slightest symptoms.

A headache might portend a brain tumor. A sudden cough, chills, or the flu were seen as harbingers of leukemia. Even the common bumps and bruises earned by outdoors roughhousing now could hint at some terrible, previously unrecognized vulnerability. Suddenly, the most durable children appeared fragile.

Every time one of Teresa Buentello's children came down with a cold, she whisked them off to the doctor. "That's how my daughter's cancer started," she remembered. "When Tresa was born, I thought I was going to have her forever. I worry a lot."

Several parents kept running tallies of the neighborhood's sick kids. Rumors spread regarding the eight year old with a large, frightening lump on his thigh. It later turned out to be a boil instead of cancer. During the winter rounds of colds and colic, parents asked one another if cancer was contagious. When a boy or girl missed more than three or four days from school, the absence aroused suspicion. Was the cancer spreading? Could a pattern be discerned? Who would be next?

Reports also circulated throughout McFarland over what now appeared as unnaturally high rates of respiratory and digestive diseases among adults. Old-timers resurrected vivid tales about former neighbors who had died years before—sudden, mysterious illnesses striking their lungs, stomach, liver, or brain. "When I first moved here in 1979," remembered Connie Rosales, "I started talking to people and they would tell me about these health problems they had in their families. I would say, 'Man, this is really a strange area.' It just seemed that these people had a lot of hard luck stories when it came to health things. You couldn't come out and say it was a problem. It was just a funny feeling I had in the beginning."

Rosales never stopped speculating about the source of McFarland's troubles. Her frankness earned her some admirers. But she also had a knack for making enemies. To many people in town, Rosales seemed a difficult woman, alternately mercurial and dogmatic.

"If these children had died from drunk drivers," admitted Rosales, "I would have been involved in MADD—Mothers Against Drunk Driving. I would have been the spokesperson, and I would have been *maniacal.* But when you're fighting against some nebulous force, it makes everything even harder. Who do you wave your sword at?"

Unlike most of her neighbors, Rosales was not Hispanic. Raised by Russian and Polish parents in San Jose, she had occasionally labored in the fields as a girl, picking potatoes and cotton alongside Spanish-speaking immigrants recruited for seasonal stoop labor. By the time she married into a Mexican-American family, she felt at home with the culture—and in McFarland, she quickly developed friendships with her Hispanic neighbors. Yet sometimes she feared that the culture's fatalism would inhibit the outrage required to propel officials into action.

"I call it the Catholic mentality," said Rosales. "This feeling that we just have to suffer. Sometimes I feel like a cheerleader, insisting, '*No*, we do not have to accept this!' "

McFarland was splitting into two factions. One side consisted of Connie Rosales, Tina Bravo, and a handful of other mothers who publicly insisted that some undetermined environmental exposure must have caused the children's cancers. The women did not form a large group, but they quickly learned how to draw attention to their cause. At meetings of the city council, school board, and water commission, they criticized McFarland's leaders, questioning their good sense, and then their good faith. Finally, they deemed the officials callous, cynical, and perversely unwilling to jeopardize the town's reputation by conducting a thorough search for the truth.

This mounting level of mistrust and anger confused McFarland's civic leaders.

"I really don't know what we could have done," admitted Rueben Garza, who had been elected to the city council shortly before the controversy erupted. "When the cancer cluster issue came into play, the city council didn't take a stand because we didn't think that the problem was necessarily limited to our town. But we did express our concern."

Actually, the city council presumed that Rosales and the others would eventually grow bored of the controversy, quiet down, and leave them alone. But they didn't. Rosales, in particular, kept hammering away, demanding action.

"Something happened here," she insisted. "I will never believe that this is 'just life.' "

At first, the town's officials could privately dismiss the mothers as "hysterical housewives." Yet publicly, they had to concede the women's right to worry. In a town with so many sick, dying, and dead children, any mother would fret. But the personal tragedies of a few families, they argued, did not mean the entire town should be alarmed. And what did any of the women know about chemicals in the water, environmental contamination, or public health?

In truth, the town officials knew no more than the mothers. And as a result, their reassurances sounded flat and unconvincing. But far more persuasive for most people than the persistent exchange of accusations and denials were the intimate details of the children's suffering—and the stories of stoicism and courage among their families.

"When Kiley was going through treatment," said Chris Price, whose daughter had contracted a cancer in her kidney, "she got real close to ET, the doll. When she didn't have any hair, ET was her friend. She said, 'Oh, he looks like me.' " Price smiled sadly at this recollection. "God's got something very wonderful in store for this child," she insisted, "after all she's gone through. This kid is great, she's my hero."

"My son went through a year of chemotherapy and radiation," recalled Connie Rosales. "I would be with him, while he was vomiting violently all night. And then all of a sudden, he would look up and say, 'Mom, I love you.' He couldn't play football. He had to take a year off, sit on the sidelines, and watch his friends play. That year, he worked out with weights in our garage all by himself. The next season, he went back and won the most valuable player award. I think he is just the most remarkable person I have ever known."

At this time, few fathers joined the public debate. Although some men thought Rosales and the other women made sense, they felt reluctant to speak out. Some feared the loss of their jobs with the local agricultural companies if they were branded as troublemakers.

"My husband doesn't like me getting involved," admitted Rose Mary Esparza, whose son had contracted cancer. "He thinks I should be home attending to my family and minding my own business. Now, I don't think so. We have to get out there and talk to people if we want to get anything accomplished."

Yet already some people whispered that the women's public tirades were hurting the town's reputation and might ruin its fragile economy.

"The god of this state is real estate," explained Rosales. "Property taxes, housing prices—they want them to go up, up, up. McFarland was just beginning to have a boom when this thing hit. We got blamed for everything that happened here. I mean, if somebody went out of business, or a house couldn't be sold—it was all our fault. It was all our fault because we brought this issue up."

Numerous people talked about leaving town. After Tresa Buentello died, the entire family relocated to Los Angeles. For months, the windows and doors of their abandoned home remained boarded up with plywood. For a time, Rosales posted a for-sale sign in her own yard. Others defiantly expressed their determination to remain.

"I won't move from here," insisted Borjas Gonzales, whose son, Franky, had his leg amputated before dying of cancer. "Not even with the risk to my other kids. I look around this house and I see so many things that Franky played with and worked on. This is Franky's house, his place. Franky grew up here, and he died here, and we're staying here."

McFarland's business leaders were concerned about the talk around town. They feared that wild speculations about the cancer cluster would drive away prospective residents and discourage invest-

Sally and Borjas Gonzales keep a photograph of their son, Franky, in the bedroom. "Franky grew up here, and he died here," said Borjas, "and we're staying here."

ment. Darrell Stinnett, who ran downtown McFarland's Western Auto Supply, complained that "since the cancer came, my business has gone straight downhill." Stinnett estimated a 30 percent decrease in sales. "We've got the image that there's a cancer problem. We've got the image that nobody cares. And McFarland has never drawn outside customers because it's got the image that there's nothing here. Those images are hard to fight."

Already the school superintendent, Ron Huebert, confided that new teachers seemed reluctant to relocate to the area.

"This isn't Newport, this isn't Sausalito to begin with," remarked a Bakersfield businessman who knew McFarland well. "Then you throw in the spectre of saying, 'Yeah, we don't know what caused these kids to get cancer, but it could be the air, the water, the soil—Lord knows what?' Do *you* want to come here?"

In the midst of this growing controversy, Rosales and several other mothers urged the county health department to undertake an investigation to pinpoint the source of the cancers. When the officials resisted, the mothers felt perplexed. By now, they had little faith in the town officers; some city council members even admitted that they didn't think they should get involved. But the mothers still expected the county professionals to rush to their assistance, eager to identify, root out, and finally eliminate the cause of the children's illnesses. The county health officials argued that in their opinion the illnesses simply did not make sense—particularly, the nine different kinds of cancers.

Eventually, one health department staffer telephoned Rosales, pointedly noting that many of McFarland's cancer kids bore Spanish surnames.

"So this doctor asks me, 'Have you been taking, you know, any of those herbs from Mexico?'" remembered Rosales. "And I said, 'Right, lay the blame on us.' He thinks we just fell off the banana truck."

This exchange hinted at the one issue McFarland's civic leaders most wanted to avoid: the town's racial divisions. Although McFarland experienced none of the violence or overt hostility that troubled many larger, more racially diverse communities in Southern California, the fact remained that the Anglo minority retained control of most public offices in this overwhelmingly Hispanic town. Anglos controlled the city council, the water board, and the public schools administration. In fact, the first Hispanic mayor was not elected until well into the 1980s.

"People of Mexican-American descent looked upon Anglo people as not caring," admitted Darrell Stinnett, who also headed the Chamber of Commerce, "not being concerned because it wasn't our kids that were dying. I felt that was very unfair. Everybody's concerned."

Many Hispanic parents talked privately about being blamed for their children's illnesses. Even a doctor from the California Department of Health Services angrily recalled overhearing local politicians tell each other that the town's real problem was the failure of Hispanic parents to bring their children in for regular physical examinations.

"If the sick ones were all blonde, blue-eyed kids," argued Teresa Buentello, "the state would have moved much faster. They would know by now what's wrong and what to do. But they just think we're a bunch of Mexicans, and we probably have it in our genes."

Connie Rosales finally decided that she might be able to pressure the local officials into action by engaging some influential outsiders. She wrote a letter to Democratic state senator Art Torres, chair of the Senate Committee on Toxics and Public Safety Management. Torres was part of the new wave of emerging Mexican-American politicians just beginning to wield power in California. Some circles even fancied him as a future candidate for governor. Rosales' letter found its way to Torres through the hands of Henry Rodriquez, state vice president of the influential Mexican-American Political Association. For the first time, McFarland's small-town troubles began to flirt with the larger and far more beguiling world of state politics.

In July 1985, State Senator Torres conducted committee hearings to focus attention on McFarland's health problems. Held in McFarland, in the shadow of the road sign along Highway 99 that welcomed visitors to "The Heartbeat of Agriculture," the meetings drew widespread publicity throughout the Central Valley and beyond—including reporters and television crews dispatched from Bakersfield, Sacramento, Los Angeles, and San Francisco.

That same month, one year after the death of Tresa Buentello, the Kern County Health Department confirmed the town's ninth childhood cancer case. With this official finding, the families who had already lost children plunged back into their anxiety and grief. Far more than the senate hearings, the new case also persuaded additional McFarland residents that the cancer cluster must have a definite cause; it couldn't just be random chance.

"I just wish that they would finally admit that there's something going on here," complained Teresa Buentello. "They should say, 'We

Teresa Buentello and her daughter, Tatiana. Tatiana's sister, Tresa, was the first McFarland child to die from cancer. Concerned about Tatiana's fate, Teresa and her family left McFarland.

don't know what it is, but we're going to evacuate the town—no matter how much it costs.' "

Local health officials began to bend to public pressure, agreeing that the cancers merited further study. The Kern County Board of Supervisors declared a public health emergency. This action, amplified by the media's attention at the Torres hearings, prompted the state to finally come to McFarland's aid. The California Department of Health Services granted the Kern County Health Department $40,000 to undertake the first stage of a comprehensive community health study. At last, it appeared to many worried parents that an answer would soon arrive.

"We were sure it was contaminated water causing it," said Rose Mary Esparza. "Chemicals. The children here call the water, 'cancer water.' "

To other people in McFarland, the sudden barrage of attention only complicated the town's problems.

Merry Ellen Alls, who taught science and math in McFarland's public schools, struggled to inculcate her students—and not incidentally, their parents—with what she believed to be a more rational perspective regarding the cancer cluster debate.

"I'm the nut in town who says there's nothing wrong," quipped Alls, who had also worked for four years at the federal Centers for Disease Control. "Scientifically, it's more reasonable to assume that it's a chance cluster situation."

When Alls' elementary school students complained about drinking the town's "cancer water," she would retaliate with a basic lesson in disease pathology, explaining that there was no evidence that the water had caused the cancers. When two of Alls' students brought gallon jugs of purified drinking water to class from home, she didn't object—until the kids started water fights in the room. "Sometimes," she admitted, "I think they did the 'cancer water' stuff more to agitate me."

But Alls also regarded the cancer cluster in terms of the tantalizing scientific questions it raised. "I remember thinking this would make a nice little research project for somebody," she recalled, "maybe a master's thesis."

So as the county officials began their study, McFarland divided neatly into two camps. On one side stood Alls and numerous other people in town who worried more about their neighbors' overreaction than additional cases of cancer. On the other side stood Rosales, Esparza, Buentello, Bravo, and many other mothers who felt certain that the drinking water, or some other environmental contaminant, had poisoned the children. What united both sides was the hope, even the belief, that the experts would soon reach a conclusive answer about their community's safety.

For all of McFarland, it appeared as though the battle was nearly over. In truth, it had barely started.

■   ■   ■

The Kern County Health Department launched its study in the spring of 1985. Dr. Thomas Lazar, a former Fulbright scholar with a Ph.D. in medical anthropology from UCLA, coordinated the research effort. At the time, no one in town questioned whether a graduate degree in medical anthropology indicated an adequate scientific background for delving into the complex mysteries of McFarland's cancer cluster.

To begin, the county team conducted a census of the childhood cancers, identifying ten sick children up to the age of 19 who had been diagnosed with cancer between 1975 and 1985. Researchers then collected water samples from homes and wells, testing them for approximately one hundred chemicals. They scraped soil from yards, parks, playgrounds, and water runoff sumps, and analyzed the samples for about eighty chemicals. The homes of the ten cancer cases—and four homes of healthy children, used as controls—were inspected for asbestos, formaldehyde, household chemicals, and other toxic exposures. County air pollution officials measured ambient levels of carbon monoxide. They even began looking at the seemingly farfetched idea that the Voice of America radio transmitter, located four miles away in Delano, had showered upon the community dangerous levels of microwave radiation.

By the study's conclusion in July 1986, the officials agreed that they could find no contaminant in the environment at high enough levels to account for the cancers. The cluster remained a mystery. Many McFarland residents felt disappointed, dismayed, and angry. Others simply felt suspicious.

"If they did know what caused it," asserted Chris Price, whose daughter had cancer, "I think it would be covered up. You're talking about an economy that's already failing. If the government admitted something was wrong, they'd have to buy all these houses in this whole town. They're going to keep testing and keep testing and keep coming up with nothing."

"We've got kids dying right and left and we need some help," agreed Connie Rosales. In the wake of the study, she announced plans to hold a house meeting in McFarland for concerned parents to draft a letter requesting assistance from the federal Centers for Disease

Control. "We have two choices here," said Rosales. "We either lie down and die, or we stand up and fight."

McFarland's mayor, Carl Boston, formed a 12-member commission to "deal with public concern and adverse publicity" resulting from the cancer cluster. The commission faced a formidable task. Even though the county had released a study finding no cause for the cancers, people continued to doubt the safety of McFarland's water supply. And in truth, the state's bureaucratic machinations had given them cause.

Around the time of the county's health study, the California state legislature had approved $400,000 for a complicated system to remove nitrates from McFarland's water supply. Many people in town read the legislature's unexpected largess as a double message about the water's safety. If the water hadn't hurt anybody, why spend nearly one-half million dollars to improve it? Public health officials strained to explain that the potential public health threat from the drinking water's high nitrate levels was unrelated to the cancers. Their arguments were sound, but many people remained unconvinced.

In this season of high skepticism, several uneasy alliances formed in order to pressure officials into further action.

In October 1986, the United Farm Workers union completed its promotional video, "The Wrath of Grapes," detailing the health risks of pesticide exposures to agricultural workers, consumers, and residents of towns such as McFarland. The video featured disturbing portraits of young children suffering from severe birth defects and cancers. On the screen, McFarland's parents unequivocally blamed pesticides for the cancers. Throughout the video, vivid scenes of pesticide-spraying helicopters whirring like gunships over the grape fields mixed with cold, lingering shots of huge pesticide containers marked with the skull-and-crossbones. Interspersed between these portraits of poison were bright pictures of young children and babies innocently devouring grapes. Mike Farrell, star of television's "M*A*S*H," narrated "The Wrath of Grapes." Connie Rosales and several other mothers of sick children from McFarland and Delano appeared on-screen.

As the camera panned high above McFarland, Rosales' voice lamented: "These homes were our dream homes, our piece of the American Dream. And it's almost like it's turned into a nightmare now. We don't know what happened here. It's out of control."

Although Rosales and the UFW seemed like natural allies, their collaboration proved shaky from the start. In fact, Rosales would soon regret her participation in the video, and "The Wrath of Grapes" would drastically exacerbate the divisions just beginning to surface within McFarland. In this small town, the terms of allegiance and enmity were growing increasingly complex and unpredictable.

By far, the greatest surprise came in January 1987, when Dr. Thomas Lazar, coordinator of Kern County's cancer cluster study, resigned from his position and publicly denounced the entire health study of McFarland as poorly designed and haphazardly executed—with its results deliberately misrepresented by the officials. Word also spread about Lazar's new job: The medical anthropologist had joined the United Farm Workers as its health consultant, promoting a new clinic to be run by the union. Soon after, the UFW dispatched fund-raising letters under Lazar's signature, pitching for donations to build the clinic. In the letters, Cesar Chavez called Lazar "a brave man" who "dared to quit a high-paying powerful job when he saw that he could not accomplish what he was supposed to be doing . . . helping the sick, and preserving good health." Others who had worked with Lazar were perplexed by his abrupt reversal.

"For some reason," mused Dr. Richard Kreutzer, a physician with the California Department of Health Services, "he now felt more like taking the shots than catching the bullets."

Over the following months, Lazar charged numerous times that county officials had downplayed the extent of the cancer problem in the area. During a speech at Fresno State University, Lazar told a chapter of the Mexican-American Political Association that he had uncovered more than three hundred cases of childhood cancer in Kern County. He characterized his findings as evidence of "an enormous public health problem." And he noted higher than normal rates of childhood cancer in the towns of Tehachapi, Edwards, Delano, and several other Kern County communities—plus a dozen more childhood cancer cases in McFarland. State Department of Health Services officials challenged Lazar to provide evidence of his discoveries. He ignored their criticism, and their entreaties.

Lazar's charges threw the investigation of McFarland's cancer cluster mystery into a state of chaos. In October 1987, state senator Art Torres chaired a second public meeting in Bakersfield of the Committee on Toxics and Public Safety Management to air the community's distress over the county health study. The committee listened to waves of angry testimony from McFarland residents, as well as other people from small Central Valley towns, who believed that their family members had contracted cancer because of some unidentified environmental contamination. Echoing Lazar's accusations, word spread about a "cover-up" and "conspiracy." Some people expressed their doubts that the county could ever complete a fair and accurate study, given its domination by agriculture. There was just too much money at stake; the discovery of a potent health hazard might devastate local farming interests.

"We should not let childhood cancer become a normal part of life!" exclaimed Connie Rosales, as she rose from the audience to speak.

At home, Borjas Gonzales, father of Franky Gonzales, simply sneered at the process that he considered a charade. "I don't think the government's got the balls to say there's a problem here," he said. "I think they're going to hide it."

Following the intense meeting, Torres demanded that a private consultant, independent of the Kern County Health Department, take over the study. Kern County officials readily agreed, urging the state to assume responsibility for the investigation. Torres' committee also expressed its intention to broaden the scope of state inquiries to several other towns suspected of high childhood cancer rates, including Fowler and nearby Delano. McFarland's problems—and fears—were spreading far beyond its borders.

Yet throughout this contentious time, many people in McFarland, as well as other residents of the Central Valley, continued to believe that the parents and politicians had vastly exaggerated their region's health problems.

"Some are guilty of the Chicken Little syndrome," complained Merlin Fagan of the California Farm Bureau Federation. "If you'd believe their allegations, everybody should be dropping dead of cancer."

Jim McFarland, a substitute teacher whose grandfather had founded the town, agreed that his community had suffered from an excess of cancer in children. But he also remained convinced that the entire controversy had been preposterously inflated by the media. "The whole idea of the town 'gripped by fear,' " he insisted, "that was manufactured. You read it in all the articles, but it wasn't happening."

Jim McFarland's criticism surprised some people. His wife's sister had died of cancer at the age of 14; health officials considered her part of the town's cancer cluster. But as a member of the town's middle-class Anglo minority—no less, a descendent of the town's founding father—Jim McFarland identified little with the Hispanic parents who led the frantic search for the cause of their children's illnesses. Rather, he spoke with more emphatic appreciation and empathy for the difficulties faced by the scientists and other health professionals who had arrived in town to study the problem. Affably dispassionate, Jim McFarland toyed with the idea that many people in town simply could not accept the injustice of cancer striking their children for no apparent reason; they had mistakenly, though quite understandably, rejected the complexity, and thus the tragedy, of everyday life. Echoing other voices in town, he even suggested that the "profit motive" drove the community's most vocal agitators, who hoped for a federal buy-out of their homes. "After a while," concluded Jim McFarland, "it all simply seemed to be an outlet for bashing the state and county officials."

In an editorial appearing in the *Delano Record*, nearby Delano's newspaper, managing editor Paul Wahl concurred that the public uproar over the cancer cluster was causing unnecessary strife in McFarland, while only adding to the confusion.

"What upsets me most about the entire situation," wrote Wahl, "is the way this extremely poignant issue is being exploited by politics and misplaced priorities." Wahl criticized local politicians whose public displays of compassion for the cancer-stricken families appeared no more than a vulgar appeal for votes. The editor also lambasted the United Farm Workers—and most fiercely, Tom Lazar. "Suddenly, Dr. Tom Lazar, the former Kern County Health Department employee who was in on much of the McFarland cancer study, is working for the UFW and grabbing the spotlight for his organization—again at the expense of the real issue."

Wahl proclaimed his aspiration to some day run one final story about the McFarland cancer cluster. He would adorn the piece with "a gigantic bold headline: McFarland cancer mystery solved— McFarland returns to normal." Many people in town felt much the same way. But all evidence indicated that a state of normalcy would not be declared in the foreseeable future.

As if to confirm the town's continuing notoriety, state assemblyman Trice Harvey announced on February 23, 1988, that he had selected Connie Rosales as "Woman of the Year" for "her deep concern and tireless efforts" in the search for a cause of the cancer cluster. "She has made a great contribution to the people of our areas," proclaimed Harvey, "and I'm sure we will be hearing a lot more from her."

Word also spread that California governor George Deukmejian had appointed a new "blue ribbon" scientific advisory panel of medical and public health experts to review the McFarland mystery. The advisory panel would offer to the Department of Health Services recommendations regarding future strategies for the investigation. This latest team of experts' sudden appearance underscored the complexity of McFarland's problems—and the mounting political pressure to find a solution.

Kiley Price, McFarland.

McFarland's reputation had spread far beyond the Central Valley, even outside the state. The *Los Angeles Times, Philadelphia Inquirer,* and *Washington Post* all ran lengthy features. The cancer cluster hit the national television news.

Rueben Garza, a McFarland city councilman who would later become the town's first Hispanic mayor, received worried telephone calls from his relatives in Texas. " 'What's this we've been seeing on TV?' they asked me," recalled Garza. " 'Why don't you guys sell everything and come on back home?' "

Around this time, Chris Price, mother of Kiley Price, who had contracted a rare kind of cancer called a neuroblastoma, received a letter from a stranger in Morgan City, Louisiana—a small town fixed upon the extraordinarily polluted petrochemical corridor of the Mississippi River. This woman's two-year-old granddaughter had also come down with a neuroblastoma. In reading about Kiley's ordeal in McFarland, the Louisiana woman had recognized her own granddaughter's suffering—and the dilemma that also now faced her own town. Parents in Morgan City, Louisiana, population 15,000, counted at least five children with neuroblastomas—a cancer cluster mystery to rival McFarland.

"She wanted specifics," remembered Price, "about how to get the health department to do something. I wrote her back and told her—*you just go all out.* You just make waves. You keep telling your story over and over and over again."

A Chemistry Lesson for Yellow Creek, Kentucky

Yellow Creek, Kentucky.

Yellow Creek winds through the hills and hollows of Kentucky, just north of the Cumberland Gap. This is Daniel Boone country, where the loose accumulation of wooden shacks and plain stucco working-class homes dot the creek's banks to compose a town that exists in name only. The nearest real town, with paved sidewalks, parking meters, and a shopping mall, is called Middlesboro, and it lies 12 miles upstream. If you don't live in Middlesboro, you're said to inhabit Yellow Creek—the designation-by-default for this portion of Appalachia's rural sprawl.

In the fall, the backwoods creek is enlivened by a flash of color in the surrounding hills. By late September, hints of yellow penetrate the forest, and a few weeks later, the thick ash and piney woods are smothered in a blanket of shocking reds from the interspersed stands of deciduous oak, buckeye, and hickory. During this season, no one talks of the trees changing color. The hills themselves turn red. Sugar maple and poplars running along the creek transform the lazy shore into a corridor of vibrant color. Yellow Creek suddenly resembles a nineteenth-century landscape painting by John Singer Sargent or Thomas Eakins—the kind of bucolic idyll in which truant boys frolic in the neighborhood swimming hole.

For four generations, Larry Wilson's family has lived about one hundred feet from the shores of Yellow Creek. Wilson's father built their newest house. After Wilson had his own children, he improved upon the structure in the ambling fashion of Appalachian design by tacking on bedrooms, storage space, and a steep unpaved driveway now ferociously guarded by the family's dogs. From either side of the house, you can spot the creek etching its sluggish path around Wilson's land.

In the 1950s, when Wilson was a boy, the creek would periodically turn jet black. Everybody in Yellow Creek knew what the problem was. Upstream, the Middlesboro tannery employed the traditional tanning method of boiling tree bark into a tea of tannin to preserve and darken leather; then it pumped the dregs into the creek. For generations, the tannery's discharges were part of the landscape—an ugly, but otherwise unremarkable discoloration of the pleasing backwoods scenery. "It was just a nuisance," confirmed Wilson. "We didn't see it as a threat."

Like many young men in Yellow Creek, Wilson departed Appalachia for military service. In 1977, he finally returned home with his wife, Sheila, and their children, to invest all their money in a small farm. But the creek had changed. All the fish were gone, except some scavenging carp. The water now stank. Vegetation along the banks had vanished. "Eventually we stopped washing with creek water because the soap wouldn't suds," recalled Wilson. "It was too greasy."

The Wilsons bought some dairy goats, feeder pigs, chickens, geese, ducks, and calves. His family planted a big vegetable garden. "We were doing great," said Wilson, "up until the summer of 1980. Then the drought hit, our well went dry, and we couldn't sustain the garden and all the livestock."

Wilson worried about the how bad the creek had become in his absence, but he needed its water to save his farm. He consulted Yellow Creek's mayor, and then talked with tannery officials in Middlesboro. The water had always looked bad, they reminded him, but it was perfectly safe.

Wilson bought a pump, threw a hose into the creek, and started watering his livestock and irrigating his vegetables. As soon as he switched to the new water supply, his pigs stopped breeding. His chickens' eggs no longer hatched. The vegetable patch withered. Goats miscarried or bore deformed kids. Every one of his calves died. Inside the Wilson house, the toilet and bathroom pipes dissolved and their stainless steel cookware turned yellow. "We stopped buying any clothes that were white," remembered Wilson. "I was ashamed to get undressed in front of anybody because it looked like my underwear had never been washed."

Larry Wilson began to haunt his neighbors' homes, "ranting and raving" about his dead livestock. Wilson cut a formidable figure, tall and chesty with the slight paunch of approaching middle-age gathered above the belt buckle of his worn blue jeans. Since he had started to go bald, he had taken to wearing a cowboy hat. People now thought of the hat as his trademark. Over time, the cowboy hat would also symbolize Wilson's new role as the community's leading rabble-rouser. Wilson was growing intensely frustrated and angry over his inability to pinpoint what had poisoned Yellow Creek.

One night he wandered over to the home of his neighbor and close friend, Gene Hurst. They shared a few beers and spun some bluegrass records. It didn't take long before Wilson was bearing down once again on the mystery of the creek.

Hurst was a quiet, patient man—in many ways, Wilson's opposite. He was careful about his appearance, dressing in neatly ironed, button-down shirts and chinos instead of Wilson's habitual old flannel shirt and blue jeans. He seemed almost ageless with chiseled good looks and a young man's sleek frame. For years, Hurst had worked as a train engineer, but his soft- spoken, deliberate manner indicated a philsophical bent. "Gene always made you think in a different way than you usually think," admitted Wilson. When guests came for dinner, Hurst's wife set the dining room table with formal place settings—a far cry from the usual quick bite at the Wilson's house amid a swarm of children on the living room floor.

When Wilson showed up at Hurst's house in a fury over the problems at his farm, Hurst was ready to listen. The water now worried him, too. And Wilson's tirade about his livestock was not the worst story he had heard. About a mile downstream, six-year-old Shannon Taylor suffered from chronic diarrhea and vomiting, dwindling to a weight of 38 pounds. Another family living along the creek had suffered six cancer deaths over five years. Other people in Yellow Creek complained about lingering bouts of diarrhea, vomiting, ulcers, kidney problems, and miscarriages. A variety of chronic ailments had always been common in this impoverished region, attributed usually to a combination of inadequate health care, poor nutrition, heavy smoking, and the occupational hazards of coal mining, such as black lung. Hurst and Wilson were familiar with these problems. But there now seemed to be an odd clustering of complaints from around the creek.

Late into their evening of beer and bluegrass, Wilson finished off his drink, glanced up at his friend's display of Indian arrowheads mounted on the living room wall, and flatly announced that he had just figured out the cause of their town's troubles. The problem was "chemicals."

"What chemicals?" asked Hurst.

"I don't know," Wilson admitted glumly, sinking back into his chair.

"Well," persisted Hurst, "how can we find out?"

Wilson suggested they learn what chemicals were used in the tanning industry.

"We both thought that made good sense," Wilson remembered. "And so I came home and I went to the encyclopedia and looked up 'Tanning.' And there I found the word 'chromium.' "

Wilson discovered that chromium is a lustrous, hard, steel-grey metallic element, resistant to tarnish and corrosion. Chromium's atomic number is 24, its atomic weight 51.996, its melting point 1,890 degrees centigrade. Wilson also learned that chromium is highly toxic and extremely dangerous to animal and human life if consumed in substantial quantities.

Armed with this nugget of information, Wilson drew together a band of neighbors who began to investigate the Middlesboro tannery. They found out that in the 1960s, the locally owned plant had been sold to a large firm headquartered in Chicago. The new owners stepped up production and replaced the traditional tanning method that had discolored the waters for nearly one hundred years with a far more efficient modern technique. This was the technique Wilson read about in his encyclopedia, and it employed chemicals suspected of causing cancer, skin diseases, and kidney ailments. Since the 1960s, the tannery had disgorged an unprecedented toxic swill of heavy metals and chemical agents directly into Yellow Creek.

"To keep people in this area powerless, you withhold knowledge," explained Wilson. "They were preying on us ignorant hillbillies."

Wilson bought a loudspeaker from Radio Shack and fastened it to the bumper of his old Plymouth Duster. Along with Hurst, he "drove up and down every holler and every road in Bell County, yelling over that thing like a bunch of idiots. '*If you want to clean up Yellow Creek, come to the school tonight at seven P.M. And bring your protest signs!*' We had hung our hat on chromium and so we started hollering, '*Chromium*—if it kills animals, it's not good for us!' We even drove down Main Street. We had no idea what would happen. We were scared to death.''

Wilson's home quickly became the local "Grand Central Station of toxic wastes." Eventually, 426 of Yellow Creek's 1,000 residents formed a new community group, calling themselves the "Yellow Creek Concerned Citizens." "We thought that was catchy. We thought we were the only concerned citizens group that had ever been formed. Later, we were astounded to find out there were hundreds of groups like us calling themselves concerned citizens."

The Yellow Creek Concerned Citizens won a court order to inspect the tannery. They documented waste disposal violations and sued both the tannery and the city of Middlesboro to stop polluting their creek. But the victory was neither quick nor painless. One night on a narrow mountain road, a mysterious car pulled alongside Wilson's Plymouth and its passenger drew a rifle bead on Wilson's face. "I saw the muzzle blast," he recalled. "The glass near my head blew out." A police investigation failed to turn up any leads. But Wilson and his neighbors were hard people to intimidate, particularly after they had armed themselves with information. "Appalachia has a long history of outside organizers," said Wilson. "The VISTA workers, the War on Poverty, the United Mine Workers, the southern labor unions—people are always coming in to save the poor Appalachians. It don't work. We have to do it for ourselves." After six contentious years, the Yellow Creek Concerned Citizens accomplished for themselves every goal that they had initially set. The creek was cleaned up.

"We couldn't match the tannery's money," said Wilson, "and we had no education to fight their science. But we had our encyclopedia."

- - -

If Larry Wilson had continued to thumb through his encyclopedia, he might have soon made another startling discovery. Only a few pages in front of the entry for chromium, Wilson would have lighted upon a far more crucial element whose extraordinary essence not only illuminates the origins of life, but also helps explain why Yellow Creek and thousands of other communities had reason to fear the modern world's unbridled profusion of synthetic chemicals.

Mary Ellen Wilson and her family live along Yellow Creek. Wilson's son was born without fingers. Every year between 1981 and 1986 a member of her family died of cancer. Her nine-month-old niece, weighing only nine pounds, also died. Wilson blames the chemicals in the creek. "How deep are these chemicals embedded in this family?" she asked.

Larry Wilson needed to turn his encyclopedia to the entry marked: carbon.

"Carbon is, in fact, a singular element," wrote Primo Levi, the Italian novelist and scientist, "it is the only element that can bind itself in long stable chains without a great expense of energy, and for life on earth (the only one we know so far) precisely long chains are required. Therefore carbon is the key element of living substance."

Every creature on the planet, from fungus to forests, from microbes to humankind, derives its organic structure from the carbon atom's malleable design. It's this commonality of basic biologic organization, our shared origins founded upon the carbon atom, that has made all life on earth susceptible to the good and ill effects produced by synthetic, petroleum-derived, carbon-based chemicals.

The systematized study of chemistry dates back 3,500 years to the ancient Egyptians. But the manner in which carbon readily conjoins with other elements to form complex chemical compounds—the basis of carbon-based, or "organic" chemistry—remained obscure for many centuries, baffling scientists until the middle of the nineteenth century. Even 150 years ago, educated people were speculating about mysterious "vital forces" that seemed to owe more to sorcery than observation and deduction.

"Organic chemistry to me appears like a primeval tropical forest full of the most remarkable things," wrote Friedrich Wohler, a leading nineteenth-century German chemist, "a dreadful endless jungle into which one dares not enter for there seems no way out."

Then in 1865, another German researcher named Friedrich von Kekule illuminated the pathway leading far beyond the jungle's reign: His map took the revolutionary form of the benzene ring.

In a sudden flash of insight, Kekule visualized how carbon atoms served as the foundation for the benzene molecule. Instead of linking together in ordinary chains, carbon looped around to forge a six-atom hexagonal ring, to which other atoms—and other benzene rings— could easily attach themselves in innumerable combinations. It was an ingenious perception that benefitted from Kekule's youthful training as an architect. Given some refinements, his benzene ring stands even today as the blueprint for virtually all important developments in organic chemistry, setting the stage for constructing the complicated synthetic compounds of the modern petrochemical age.

The structure of benzene—found in petroleum, and thus derived from living matter—also explains why nearly all of the six million natural and synthetic compounds that contain the benzene ring may have an effect on living beings. The benzene ring forms the backbone of thousands of chemicals that orchestrate the daily events of our lives. For example, the natural sex hormones, estrogen and testosterone, are a series of linked hexagonal benzene rings combined into a structure that resembles chicken wire with hydrogen atoms attached. Cholesterol is a similar chicken wire jumble of benzene, as is the stress hormone, cortisone. So when benzene is chemically recycled into plastics, pharmaceuticals, food preservatives, or dyes, living beings often recognize the familiar benzene structure and interact with it as a component of life itself. Scientists know little about how most of these interactions take place, though some of the human body's more apparent accommodations are quite alarming. Benzene-derived synthetic dyes, like benzidine, have been known for decades to cause cancer. Pure benzene itself is a cause of leukemia.

Prior to Kekule's discovery, scientists who wished to replicate in their laboratories the chemistry of living matter were faced with a task similar to undertaking a complicated feat of engineering—say, building a skyscraper—without ever having seen the object; without even consulting the architect's plans. The benzene ring revealed the necessary shape of organic compounds. And it provided a model for limitless tinkering.

By the late nineteenth century, the benzene ring had spurred on key discoveries in the coal-based German chemical industry. For nearly one hundred years, German research dominated the marketplace, pioneering an array of synthetic dyes, fertilizers, and powerful new explosives. When the German chemist Fritz Haber received the Nobel Prize in 1918, he stated in his acceptance speech that so many fundamental secrets had been uncovered by the burgeoning industry that "the days of complex chemistry were over." In fact, they had just begun.

With the help of catalytic cracking—a method for decomposing crude oil and gas under intense heat and pressure—chemists were able to manipulate the rings, chains, and branches that bind together multiple carbon atoms to forge inventive new configurations with other elements. In fact, the process proved so easy, and the potential benefits and profits so great, that the temptation to tinker without thought of future consequences became impossible to resist.

There was also another good reason to experiment fast and furiously with new chemical compounds. As World War II approached, Germany still led the world in synthetic rubber research, a key factor in military as well as industrial might. Once the Japanese cut off natural rubber supplies in Asia, Allied chemists had to turn to natural gas and petroleum stocks to devise a synthetic substitute. In the process, the United States spawned the modern petrochemical industry.

In fact, the Allied forces could not have waged war, never mind achieve victory, without the persistent innovation of the American chemical industry. The production of synthetic rubber, powerful new solvents, chlorine-based poison gases, and lightweight metal alloys for aircraft production gave absolute credance to Dow Chemical's new wartime slogan: "Chemicals Indispensable to Industry and Victory." But this new emphasis in modern warfare also drew into focus the dark side of synthetic organic chemicals. German chemists aided the Nazi's program of genocide by manufacturing Zyklon-B, the poisonous gas that exterminated millions in the concentration camps. And IG Farben, Germany's leading chemical company, had also operated its own concentration camp, using slave labor to meet wartime production quotas. At IG Farben's camp in Monowitz, over 25,000 inmates were worked to death. Whatever horrors the Axis' and Allies' chemical industries had produced in pursuit of victory, the rehabilitation of wartime chemical engineering proceeded briskly at home, as a booming post-war economy heaped new demands upon the nation's research labs. After the war, thousands of scientists throughout the United States and Europe were employed to bind together chemicals in countless ways to meet peacetime production needs.

Starting out with Kekule's benzene ring, these researchers added a chlorine atom here, a methyl group there, and *presto:* They found themselves gloating over something entirely new under the sun whose practical applications—and biological effects—might not be understood for decades.

Today it is difficult to overestimate the importance of the chemical industry in modern life. "Not only do chemicals permeate our homes, our clothing, our food, our cars, our offices," noted one chemical industry historian. "They *are* our homes, our clothing, our food, our cars, our offices." Since the end of World War II, the world's chemical wizards have reframed the deadly, destructive agents of warfare into undreamed of benefits for the common person. Highly effective solvents originally used to degrease wartime fighter plane engines soon revolutionized heavy industry, pushing American firms to the top of the world market. New plastics and metal alloys designed for lightweight mobile weapons systems took on startling new shapes: office furniture, tract home building materials, automobile parts, children's toys. Synthetic fibers once cut to the shape of tents, parachutes, and uniforms set the standard for civilian fashions. And the human-lifesaving insecticides that wiped out typhus-carrying body lice and malaria-spreading mosquitoes in Naples and on the beaches of Saipan were generously sprayed upon hundreds of thousands of suburban gardens beset by a modest invasion of aphids or black flies.

From 1923 to 1950, while the United States' industrial capacity doubled, the chemical industry's activity increased nearly fivefold. Production climbed from 2 billion pounds in 1940, to 30 billion in 1950, to 300 billion in 1976. Research also increased the commercial demand for heavy metals as the nondegradable, highly toxic hand-maidens of the sprawling new petrochemical industry. Worldwide production of mercury has doubled since World War II; lead production is doubling every 20 years, along with substantial increases in the industrial applications of cadmium, beryllium, and selenium.

As a result, the world has been showered over the past 50 years with some 65,000 new chemical compounds, each theoretically capable of affecting living creatures in totally unpredictable ways. These substances have literally spread *everywhere:* under the sea, atop uninhabited mountain peaks, in regions as remote as Antarctica and the Sahara desert. These chemicals also turn up regularly in the living tissue of animals, including human beings. This abrupt profusion of chemicals—and their organic link with the earth's living creatures— is the reason why Larry Wilson of Yellow Creek, Kentucky, and thousands of other people throughout America's toxic towns have come to look upon the chemical age with growing alarm. No species on the planet has yet had the time to adapt to the thousands of new chemical compounds; biological accommodation depends upon the slow and uncertain process of evolution, taking at least thousands of years. Given the fact that all organic chemical compounds have the potential to alter—that is, damage or destroy—the routine functioning of organic systems, the ceaseless worldwide proliferation of synthetic chemicals raises some chilling questions about global public health.

Unfortunately, nobody can answer these questions. The long-term biologic effect produced by most of the synthetic chemicals now blanketing the earth is unknown. Their combined effect is incalculable. And as a result, every plant, insect, fish, animal, or human being alive today is swimming blindfolded in a vast chemical ocean of unpredictable consequence.

Of course, some of the new chemicals developed by scientists were *intended* to have a biologic effect. After all, the operative syllable in the word *pesticide* is *cide*, Latin for "kill." And as a class, pesticides serve as a good example of how the effects of even the strongest biological agents may be masked for decades.

Take the case of DDT, the most notorious pesticide of the postwar years. The chemical compound DDT was first synthesized by a German chemist in 1874. But DDT's lethal potential as a insecticide wasn't appreciated until 65 years later. That's because DDT's creator wasn't searching for a poison; he was simply exploring, linking together complex molecules in the spirit of pure science, because it was suddenly possible to do so in the wake of Kekule's discovery. In the 1940s, after DDT was found to kill insects, it was applied to millions of infested farmland acres and sprayed above countless waterways to rid communities of disease-spreading mosquitos. But DDT was also having unintended lethal effects on a wide range of innocent bystanders, including birds, fish, and other wildlife, who had not been targeted for eradication, but had picked up the pesticide anyway through the food chain. By the 1960s, DDT was found in the tissue samples of human beings. In 1972, the pesticide was finally banned in the United States. The damage it had wrought, as well as the good it had achieved, made it a potent symbol for the perils and complexities of the new chemical age.

Unlike most pesticides, the vast majority of the chemical industry's 65,000 new compounds, were *not* created with the intention of affecting plant, animal, or human life. But good intentions do not necessarily protect living creatures from their insidious effect. Due to their organic basis, many synthetic compounds now in common use may slip far beyond the contrivances of their makers, spawning a great range of human harm.

The distance between chemical peril and safety is often slight. In many cases, it takes only a minor change in molecular structure to render an otherwise benign substance into a potent carcinogen. The noncarcinogenic compound anthracene can be transformed into a weak carcinogen with the addition of a single atom; the substitution of two atoms forms a very powerful carcinogen. Other subtle structural changes obscure the potentially lethal force of widely used commercial compounds. BCME is a solvent used to manufacture water repellents for clothing. It's formed from ether, with a salt molecule attached. But the chemical composition of BCME differs by only one molecule from nitrogen mustard, the deadly "mustard gas" used in trench warfare during World War I.

By reshaping compounds, scientists have also found that they may inadvertently create chemicals that are not only dangerous, but also highly variable in their specific impact. The most subtle changes in the structure of a carcinogenic nitrosamine can change the location of the disease it spawns: Add a small molecule called a methyl group to the larger nitrosamine molecule and it causes cancer of the esophagus; add another methyl group and an ethyl group—merely one atom different than the methyl group—and the nitrosamine causes bladder cancer; and add two butyl molecules and it will lead to liver cancer. Similarly, the dosage pattern of the compound can alter its effect: Low doses of dimethyl nitrosamine over a long period of time can cause liver cancer. Higher doses over a short period may produce kidney cancer.

Adding yet another layer of complexity are the uncharted effects of chemical mixtures. Given the almost infinite number of combinations among the world's 65,000 new synthetic chemicals—all at an incalculable range of doses and lengths of exposure—it's easy to see why scientific inquiry would be quickly swamped by the possibilities.

For Larry Wilson, in Yellow Creek, Kentucky, and thousands of other worried people in contaminated communities throughout the nation, this universe of unanswered, perhaps unanswerable questions has caused great anxiety. And the more these ordinary citizens learn—starting from the day they open their encyclopedias and begin to formulate their own questions—the more they feel inclined to take matters into their own hands. This impulse to direct action is not new.

Public debate regarding society's potential chemical contamination can be traced back to nineteenth-century England's alkali works. Alkali was the first synthetic substance to be widely produced for manufacturing purposes. It halved the melting point of sand in the manufacture of glass, and produced a soft paste that formed a durable cake soap when treated with lime. By the late 1700s, soap manufacturers and glassmakers promised a fortune to whomever could replace natural alkali vegetable sources with an inexpensive synthetic substitute. The alkali works depended upon manufacture of sulphate of potash, which involved using a "clear, colourless, fuming, poisonous, highly acidic, aqueous solution of hydrogen chloride."

The manufacture of synthetic alkali began in the 1820s. Forty years later, England's alkali works employed 19,000 men who produced finished goods valued at £2.5 million annually. In 1878, the governmental Commission on Noxious Vapours reported that "one or two (alkali) works hastily constructed" had "carried on in the most reckless fashion," quite probably killing all the fruit trees in one complainant's locality. Elsewhere the works' sulphur waste products—emitting a rotten egg smell—were released under the cover of night, spreading "a pall over the countryside around the factories." As years passed, the progress of the alkali industry was indelibly recorded upon the English countryside.

In the words of one historian, rural England underwent a "melancholoy transition to the drab lifeless grey of industrial wasteland." A "forest of chimneys" belched at least 13,000 tons of commercial acid over northern England to be spread "at the whim of the prevailing winds." According to contemporary descriptions of the countryside where alkali works were situated, the landscape resembled "a World War One Flanders battlefield, with once luxuriant trees browned and bare and the fields taken over by the stunted elder bushes or deadly nightshade. This was Blake's vision of England's once green and pleasant land in the grip of dark, Satanic forces with a vengeance!"

Alkali producers, vital to the national economy, like the petrochemical industry of today, were finally regulated due to the mounting public outcry. But also like today's EPA, the regulatory commission was critically understaffed and fundamentally divided as to its true mission. In fact, government regulation seemed to have been established chiefly to ward off citizen demands for more draconian measures—or worse, spontaneous vigilanteeism undertaken by enraged citizens.

"Certainly the protection of private property, rather than the protection of the nation's health," concluded the British scholar Anthony Wohl, "was the underlying motive behind the early investigation of the alkali industry." As the Commission on Noxious Vapours keenly advised: It was "obviously the duty of the legislature to be very cautious in dealing with a trade which (employed) so large a portion of the manufacturing industry.''

Alkali production wasn't the only industry to leave its mark on Victorian society. In 1861, a national health survey reported that "the canker of industrial disease gnaws at the very root of our national strength.'' The widespread and unregulated use of chemical agents spawned numerous health problems among England's new working class.

"One did not have to possess the penetrating acumen of a Sherlock Holmes to detect the specific trade of various working men,'' wrote Wohl. Workers' roles were betrayed by the chemical taints induced from their daily labors: "the potter by his asthma, the lead worker by his paralyzed wrist, the brass worker and copper worker by the tell-tale greenish tint of the hair, teeth and clothing, the pottery and earthenware worker by his blue gums, the worker handling mercury by his 'trembles,' and the confectionery worker by the skin boils which were the first sign of arsenical poisoning.''

It would be comforting to think of these workplace horrors as grim relics of the distant past. Adolescent chimney sweeps contracting "soot wart,'' or cancer of the scrotum, from their daily exposure to coal tar, or phosphorus-laden match workers developing "slack jaw,'' in which teeth loosened and the entire jawbone slowly dissolved, seem like barbaric anachronisms that could not have survived beyond the convulsions of early industrialization. Yet even today, if not especially today, the persistent exposure to highly volatile petrochemicals has created its own array of occupational diseases. Modern dye workers suffer from high rates of bladder cancer, due to their exposure to benzidine. High lung cancer rates plague workers handling nickel and chromium. Workers exposed to benzene have increased rates of leukemia. And it's only recently that some states have required those awkward rubber guards to be installed on fuel tank nozzles to shield consumers from benzene exposure during a routine fill-up at the gas station.

Of course, Victorian workers had no reason to believe that their health was being safeguarded by either government or industry. But contemporary society does vouchsafe this protection. And the promise of security extends beyond the workplace. Today the twin guarantees of industrial expertise and government oversight claim to erect an impenetrable wall between the chemical world's lethal threat and the general public's health. Unfortunately, the record on public safety continues to reflect the same attitude of malign neglect that characterized the earliest phases of chemical manufacturing.

"Dangerous products, accidents at chemical plants, and hazardous waste have always been an integral part of the industrial processes used and the way firms approach their products," asserts Lee Niedringhaus Davis, a chemical industry historian. "What is different today is not the cause or nature of these problems, but their uncontrolled momentum and size."

Like the alkali-saturated residents of England's once-verdant fields that were burned brown into chemical wastelands 150 years ago, the people in America's contaminated communities have seen their lives plunged into scenes of stunning disorder.

- - -

In truth, the real story behind America's toxic encounter cannot be told through historical anecdote, official surveys, or EPA statistics. Rather, the most instructive lessons regarding North America's immersion into a man-made ocean of poisonous waste can be found in the mounting number of human dramas played out by ordinary people whose lives have been changed overnight.

In Columbia, Mississippi, neighborhood kids slipped through an unguarded waste dump and turned the area into a playground. When scientists arrived to test the soil's toxic levels, they wore protective "moonsuits," covering their bodies from head to toe, while cautiously plodding through the same area where the children had romped barefoot. The news that their kids had been exposed to a toxic playground came as a complete surprise to most parents.

In Pilcher, Oklahoma, abandoned copper, lead, and zinc mines overflowed during heavy rains, bubbling up from "a 10 billion gallon vat of subterranean poison" large enough to flood the city several times over. The extruded acid water burned grasslands, scalded horses' hooves, and ate through a five-gallon metal bucket. "Fish had open sores," recalled one Pilcher resident, "like somebody took a knife and cut a chunk out of them."

In Jersey City, New Jersey, three million tons of chromium slag—one of the most potent known carcinogens—was used for 25 years as landfill under the sites of new homes, factories, parks, a trucking depot, and a school. A drive-in movie was opened atop the world's largest chromium dump—one million tons of slag. "The drive-in wasn't paved and the cars driving around would have easily kicked up the chromium dust into the air," said a New Jersey state health official. "Years from now, we may find people who did an innocent thing like going to the movies developing cancer."

In Toone, Tennessee, several farming families spent four years in what seemed to be chronic depression, until they discovered chloroform in their drinking water at levels equaling a daily sedative.

"I just didn't feel like doing anything," remembered J. D. Boyd. "I didn't even have the energy to get up and go to the table to eat. There were times I was on my way someplace and I'd lose track of where I was at. I'd be lost. The whole family was that way."

The well on Boyd's property had been contaminated from illegal dumping by the Velsicol chemical company. The entire community had been poisoned.

"The water would look greasy," said Boyd's neighbor, Steve Sterling. "You'd bathe in it and you'd get a burning itch. Some of us broke out with little welts, like we had chicken pox or something. A while after the bath, it'd go away. Then you took another bath and you had the chicken pox again."

The truck driver who hauled Velsicol's leaking barrels of chemical wastes from Memphis to Toone lost his sight after continued exposure to the corrosive fumes. His 29-year-old son, who worked along-

In Toone, Tennessee, chemically tainted water incapacitated J. D. and Ethel Boyd. The contamination created such concern that towns nearby posted road signs touting the safety of their own water supply.

side him, developed nerve damage. Although formerly a keen athlete, the young man's once-powerful legs began to tremor uncontrollably; one summer, he drowned while swimming.

"I wouldn't have thought no person would have done it, this dumping," said Sterling. "It's about the same thing as coming up and shooting you in the back. All of my family—wounded 'em for life."

In Tucson, Arizona, the drinking water was inundated with an airplane engine degreaser called TCE. This solvent has been linked to cancer, central nervous system disorders, liver and kidney damage, blood diseases, and lung, respiratory, and cardiovascular problems. One teaspoon of TCE can contaminate 250,000 gallons of water. Over 30 years, the U.S. Air Force, Hughes Aircraft, and the local airport dumped into ditches and unlined troughs four thousand gallons of TCE that eventually seeped into Tucson's drinking water.

"Around here, we're always talking about a funeral or an illness," said Rose Marie Augustine, a 53-year-old resident of Tucson's Southside neighborhood, where the contamination is most intensely concentrated. "This community is filled with death and illness." Augustine has contracted lupus, a progressively debilitating disease of the immune system. Her daughter had cancer. Her son was recently stricken with a disease similar to muscular dystrophy. "Muscular dystrophy is supposed to be passed on to you from your father. But for five generations, nobody in my family had it. A genetics doctor told me that somewhere along the line, there had been a genetic breakdown." Of course, Augustine can never know whether her family's illnesses were chance events or the results of chemical poisoning. And this uncertainty only added to her pain.

In Nitro, West Virginia, Linda King and her husband bundled up their three children and their belongings thinking that they could flee to a safer place by relocating westward. They had grown desperately tired of the industrial region's notoriously corrosive air, the skies slabbered with black, blue, and yellow swatches of chemical effluence. Far worse than Nitro's suffocating air was the family's continuing health problems.

Linda King's youngest son suffered from a barrage of allergies and began to lose his hearing. Her husband's hair "fell out by the handfuls." King sometimes found herself gasping for breath. For two weeks at a time, her skin broke out into a mass of red dots and her joints swelled. "In Nitro, I entered into a world I never knew existed in the United States," said Linda King. "I felt defenseless to help my family. I was an ordinary housewife. I didn't know what toxic waste was, until it turned up in my own backyard."

When the King family moved to Harvey, Louisiana, they found that the same kind of problems were even more pervasive throughout their new state. Louisiana led the nation in the amount of toxics released into the water each year. Even the lake facing the governor's mansion was so polluted that signs had to be posted advising against drinking, swimming, and fishing. Further west, on the Calcasieu River, where locals claim that the water smells like "nail polish remover," state health officials had warned residents not to eat the water's contaminated trout. But intense pressure from local industries quickly led to an astonishing revision: Soon the health department was promoting special recipes "to minimize the risk" for anyone who did plan to devour their catch.

"I'd like to know how in the world they expect you to cook a contaminated fish?" wondered Louisiana's popular Cajun chef, Justin Wilson. "You can throw away the broth"—as the state officials suggested—"but you still got the fish."

Throughout the country, more than a dozen communities had to be permanently evacuated by the state and federal governments because of untreatable toxic threats. Other communities left on their own. In Brookhurst, Wyoming, residents abandoned their chemically contaminated homes, turning the neighborhood into a ghost town and leaving their houses marked with anti-toxics graffiti. In Ponca City, Oklahoma, blue-collar families were paid $40,000 per home to leave their neighborhood; the Conoco oil company, which had contaminated the community's drinking water, planned to buffer its refinery with a zone of deserted houses. In Oxnard, California, Linda Paxton moved away from her two-story duplex after learning that it had been built upon a field of underground liquid toxic waste pools; she painted one side of her abandoned home with a huge skull-and-crossbones, announcing: *Oxnard's Love Canal. Our Home Toxic Dump.* Her neighbor, Steve Blanchard, agonized for months over the fate of his own house. Blanchard had a young daughter, and he worried about the possible risks to her health. He wrote his bank, informing them that he would no longer pay the mortgage on the contaminated property and he left.

"To walk away from everything I worked for over years and years," proclaimed Blanchard. "That's an awesome feeling. I had had all this hope. Now I feel like I might lose everything."

These disconcerting images of ordinary people confronting chemical contamination run on and on, touching every state, affecting hundreds of thousands of lives—even reaching people on both sides of the international border.

Just south of Palm Springs, California, a golfer watched the New River at the fashionable Del Rio
Country Club. Health inspectors traced at least one hundred toxic chemicals to U.S. and Mexican
factories just south of the border. The water was so contaminated it foamed as it flowed past
the golf course.

Every day in the Mexican town of Matamoros, just below Texas,
more than ninety *maquiladores*—American-owned factories moved
south for the cheap labor and lax environmental regulations—pump
an enormous amount of chemical pollution into the skies and waters.
Just over the border in the South Texas Rio Grande Valley, eighty
babies were born with fatally undeveloped brains during a five-year
period—more than double the national average.

In Rednersville, Ontario, the Canadian Environment Ministry
had to provide free bottled water and household water filter systems
to the entire town—after digging up 180 illegally dumped drums of
toxic waste generated by the Goodyear corporation. Twenty different
solvents, including cancer-causing benzene, had leaked into the
town's water supply. A government investigator called it "one of the
worst cases of soil and groundwater contamination I've ever seen." In
Montreal, the *Societe pour Vaincre la Pollution* reported in 1991 that the
firm hired to clean up the South Shore's tainted groundwater had
continued to dump toxic sludge. Along the Ste. Clair River in Ontario,
researchers discovered a huge, glutinous blob of cancer-causing

chemicals, presumably pumped into the waters by industries operating in both Canada and the United States.

In Pompano Beach, Florida, tiny white particles of vinyl chloride descended from the skies like snow, covering parked cars and swimming pools and sticking to screen doors. This extremely powerful carcinogen was being pumped into the air from a nearby polyvinyl chloride pipe factory. A health study organized by the local residents found high incidences of liver, heart, and kidney disease and cancer. In Pearland, Texas, the chances of contracting cancer are calculated as being the highest in the nation—with industrial pollution factored as a leading cause. Yet when a local housewife called the nearby landfill "a dump," she was sued for slander by the operator and her husband was threatened with a suit for "failing to control his wife." In Springfield, Vermont, the residents of a mobile home park didn't realize they were sitting on a vast wasteland of arsenic, benzene, cyanide, and lead—what the regional EPA administrator called "a chemical time bomb" —until "there were people in white suits and rubber gloves and rubber boots walking around" testing the soil.

"I remember growing up in the 1950s," recalled Bob Aufiero, whose daugher, Jessica, contracted cancer in a contaminated neigh-

Jessica Aufiero is one of two children who survived a cluster of twelve leukemia cases in Woburn, Massachusetts, where the water was chemically contaminated.

borhood of Woburn, Massachusetts. "We used to see these films about the Strategic Air Command and Sky King—like there was a big safety net over this country. The government was watching for attacks, industry was making the country strong." Aufiero wagged his head in disbelief. "My confidence has been shaken," he said, "right down the line."

Sally Teets and the Small-Town Toxic Rebels

Spencerville, Ohio.

$\mathbf{O}$n Valentine's Day, the house-cleaner finally asked Sally Teets if she could take down the Christmas tree.

"No," Sally answered morosely. "Just put some hearts on it and leave the damn thing up."

Over the past four months, the losses had piled up so high around Sally Teets that most days she couldn't contemplate any plan beyond breakfast. At the end of October, her husband's medical partner resigned from their thriving practice and left town for good. Two weeks earlier, her mother had died, following a long struggle with cancer. Most traumatic of all: On November 15, 1987, Sally Teets, chief tactician and rabble-rouser for the Dumpbusters of Spencerville, Ohio, triumphed in her 11-month battle against a five-hundred-acre landfill slated for her small town by Waste Management, Inc.

"I was never more depressed in my life," admitted Sally.

The news of her community's victory arrived in the worst possible way. Sally had spent the day in Columbus, explaining to Senator John Glenn's staff the complexities and political significance of Ohio's grassroots environmental movement. This meeting marked the enormous changes that had affected Sally's life in less than one year. After leading her own town's fight against the nation's largest waste disposal company, she now understood the toxic rebellion breaking out across small-town America in ways the politicians never would. When she returned home, the answering machine blared a half-dozen familiar, slightly hysterical voices, crying and laughing between the electronic bleeps on the scratchy, overused tape: *The fight is over, we won!* A few days later, the state EPA dispatched an official to Spencerville to order Waste Management out. Ecstatic and triumphant, the Dumpbusters closed off the city streets, roasted three hogs donated by local farmers, and threw a feast and street dance for 2,500 Spencerville residents and their supporters throughout the region.

All the while, Sally kept asking herself: *How dare this be over now?*

Until this moment of victory, Sally Teets hadn't really grasped how important the fight had become to her. Winning now, so suddenly, meant an immediate, cold-turkey withdrawal into private life. She felt deeply depressed by the mere thought of returning to the old routine: the obligations of Junior Service League and the physicians' wives' Medical Auxiliary; the afternoons spent ferrying her three children around town for their after-school music lessons and sports; the long letters composed for her college friends, detailing the minutiae of rural life; the strained cocktail parties and pretentious formal dinners that never came off quite right, but loomed over the next several decades as the unavoidable fate of small-town strivers; the laundry, the meals, the housework.

Reduced to all this, when for 11 months, she had been . . . *a warrior.*

Like the men of her father's generation who reminisced about their soldiering in Europe and the Pacific, Sally would forever regard these fighting days as the best of her life. Even now she cherished a private knowledge that the war against Waste Management had forged her into a different person—braver, more cunning and determined, steady in purpose and inspired by what she had to admit was a taste for blood.

In the year of the dump, Sally Teets discovered her genius for leadership.

The conflict began when Waste Management, Inc., proposed crowning a huge open field in the midst of Spencerville with an 125-foot-high pile of trash to be gradually accumulated over the years from several neighboring states. When Ohio's governmental regulators tentatively approved the plan, many Spencerville residents felt that one final, intolerable insult had been heaped upon the community.

Sally had lived in Spencerville only since 1981, but she understood her neighbors' anger. About ten years before, the world's first transcontinental hot air balloon had crashed about a mile from Spencerville. The national news outlets never mentioned the town, instead identifying the accident location as "the Lima area." Spencerville had always felt itself cast in the shadow of this larger city. Lima even committed the ultimate small-town insult of ignoring Spencerville's high school sports in its countywide newspaper. To many people in Spencerville, the outside world seemed to regard their community with calculated indifference, if not contempt. The dump proposal proved it.

Sally Teets in Spencerville. The field behind her was the designated site for the five-hundred acre, 120-foot tall dump. "The pile of garbage would have towered over every barn and grain silo in the area," said Teets.

Sally cultivated these ancient grievances. Along with a handful of other mothers, she formed the Dumpbusters to alert Spencerville to the proposed dump's risks. The Dumpbusters dwelled on the impossibility of separating toxics from ordinary garbage—chiding that a non-hazardous landfill taking in wastes from one hundred thousand people also inevitably accepted about five hundred tons of hazardous wastes each year. They decried the traffic of 350 trash-filled trucks that would rumble into town each day. They harped on the insult to their community, stressing the unfairness of Spencerville absorbing the refuse from communities as far away as New England.

In a matter of weeks, the Dumpbusters changed the face of Spencerville. At Sunday afternoon pizza parties, dozens of families fashioned hundreds of Dumpbuster insignias—an encircled dumptruck with a slash through the middle—posting them in the living room windows of every local supporter. Work crews constructed front lawn placards: *Give a Hoot, Don't Pollute. Help Keep America Looking Good. Please Leave Me a Green and Peaceful Planet.* At the edge of town, they erected an official-looking road sign: *Welcome to Spencerville, Home of the Dumpbusters.* The mothers, along with their children's brigade, the Junior Dumpbusters, tied three thousand red ribbons "around everything that stood still. Trees, lampposts, everything." As months passed and the organization gained more strength, Sally recruited Spencerville's most prominent citizens to serve as the Dumpbusters' official leaders. Publicly, the school superintendent and a long-time resident who headed up the Girl Scouts now spoke for the Dumpbusters. Sally Teets receded into the background, devising new methods of making life miserable for Waste Management executives.

From the beginning, Sally realized that she couldn't ask most of her neighbors to picket, demonstrate, orate, or otherwise make public spectacles of themselves. Spencerville was a typical middle-American small town, not inclined to protest government decisions or question authority. "Instead, we drew on people's special talents," explained Sally. "If somebody was a good artist, I'd ask, 'Would you do caricatures for our meeting agenda?' *'Oh, you're really important in the church!* Would you work with the churches and get them involved?' *'You're a school teacher!* Would you do something in the schools so the kids will go home and nag their parents about recycling their aluminum cans?' "

One man in town volunteered to create a slide show, complete with background music and a dissolve unit that blended one image of the town into another. The presentation commenced with Kenny G. performing "Songbird" as the screen melted into successive portraits of Spencerville at its bucolic best—its narrow main street rolling through the middle of town, the storybook-perfect clapboard homes, the children trailing home from school along quiet, tree-lined streets. Then the "Ghostbusters" theme kicked in, as garish shots of garbage-heaped landfills flashed across the screen. Near its conclusion, the show presented Spencerville residents painting signs and wrapping the town in red ribbons, as "Nothing Can Stop Us Now" blared in the background. The finale featured country singer Lee Greenwood's rendition of "God Bless America."

The slide show, like most of the Dumpbusters' antics, invoked Spencerville's latent patriotism—its fierce, sentimental vision of an America where ordinary people stood together and solved their own problems. The Dumpbusters showed Spencerville residents their country at its best.

After a few months, the fight against the dump infected every aspect of life in Spencerville, giving the town an extraordinary boost of energy. A typical weekend ran something like this: Friday night, the VFW dance, proceeds going to the Dumpbusters; Saturday afternoon, the Boy Scouts paper drive and chicken fry, proceeds to Dumpbusters; Sunday evening, the church ice cream social, proceeds to Dumpbusters. The following weekend might feature a bake sale or sidewalk sale days, proceeds allocated, as usual, to the Dumpbusters. With this money, the group paid for its supplies, employed a geologist as a technical consultant, and hired a lawyer when it was sued by Waste Management.

" 'What the hell are they going to do next?' That's what people were asking themselves," remembered Sally. "There wasn't anything else going on in town, so people said, 'Let's go see what the Dumpbusters are doing.' For 11 months, Spencerville was a wonderful place to live."

The Dumpbusters raised $137,000, an extraordinary sum for a community organization in its infancy. For $100, donors received a Dumpbusters hooded sweatshirt. Some households ordered three. But the single most-successful fund-raising event surpassed all expectations for homespun wit and audacity.

On the Halloween weekend, Sally and the Dumpbusters laid out a huge grid in the middle of the high school football practice field—four hundred squares, each square measuring three feet by three. They fenced the perimeter with bailing wire and drove posts into the ground at each corner. At the center, they plastered a big, black X, turning the design into a gigantic bingo card. The theme of this game: *We're Tired of Being Crapped On.*

Players purchased squares for $10 each. Then a cow wandered out onto the grid. "The first place the cow crapped," explained Sally, "if you owned that square, you won $500. The second crap, you won $300.

The third, $200. We had 396 squares at $10 a square—that's $3,960." Most of the winners returned their prize money to the Dumpbusters. In an hour and a half, the group made over $4,000.

Since the press in Lima still ignored Spencerville even during its finest hour, Sally recruited the local rock 'n' roll stations' most popular disc jockeys, dressed them up in tuxedos, and invited them to the crap-out as celebrity judges. Throughout the week, the deejays plugged the event over the airwaves, donating a fortune in publicity and drawing hundreds of people to Spencerville. Sally's husband hauled an old baby scale out of his doctor's office, in case the excrement landed in two squares and required judicious weighing. Volunteers wielded gold-painted shovels to scrape the grid clean. Some Spencerville gamblers brought along their fishing poles, tied an ear of corn to the line, and attempted to lure the cow to their squares.

Later that night, the Dumpbusters threw a costume party with a topical theme. Spencerville voters had recently approved a new zoning law designed to exclude the dump, which in turn prompted Waste Management to challenge the ordinance in court. The costume party finessed this problem with another incitement to patriotism. "Here it is," Sally lectured her neighbors, "the bicentennial of our Constitution, and we're being told by a local polluter that we can't even vote to determine our own welfare." The citizens of Spencerville thronged into the high school gymnasium attired in floor-length gowns, high-collared colonial suits, and powdered wigs, like George and Martha Washington. They draped a huge American flag across the furthest wall, and below it, the POW/MIA flag. Red, white, and blue balloons floated through the air. The public address system rattled "The Battle Hymn of the Republic," "America the Beautiful," and "Bless the Beasts and the Children."

Two weeks later, the state succumbed to the Dumpbusters' incessant pressure, and denied Waste Management's permit. Sally called Will Collette, her organizing mentor at Citizen's Clearinghouse for Hazardous Wastes, to tell him the good news.

"Oh, my God, Will! We won!"

But her voice sounded strangely reserved, almost melancholy. Will Collette was a professional community organizer who knew that victories of the kind won in Spencerville never came easy. "Aren't you excited?" he wondered.

Sally felt more like crying. The battle of Spencerville had ended—taking with it the energy, inventiveness, and camaraderie that had brightened Sally's life for nearly a year. From then on, she marked time from "the day I lost the fight"—meaning the day the Dumpbusters won.

"Sally," Will told her, "there are other fights out there."

Three weeks later, Sally telephoned Will once again.

"You've got to give me a title so that I can speak on behalf of Citizen's Clearinghouse," she told him. "I don't want to be out here as a lone dog. I want to be part of something bigger."

Almost one year later, the Gund Foundation in Cleveland offered Citizen's Clearinghouse funding to hire a Midwest Regional Organizer. Sally accepted the job, taking on 14 states for an annual salary of $16,000. As a roving organizer, paid to travel across toxic America and stir up as much trouble as possible, her life began again.

- - -

"I like the idea," boasted Sally, "that I handle 14 states from my home in Spencerville, Ohio. I like that it takes me an hour and a half to get to the airport. I can tell people, 'I live in a community of two thousand. I know what you're up against.' "

As the Midwest Organizer for Citizen's Clearinghouse, Sally maintained an exhausting schedule, spending half her life on the road. When we met her at home in Spencerville, she had recently returned from "a toxic tour of Nebraska, hitting seven towns in seven nights." Throughout the month, she had traveled 15 out of 31 days. The month before, she spent an equal amount of time consulting with grassroots groups spread across a region so generously defined as midwestern that Montana fit into its furthest corner. In all, she worked with over five hundred community organizations, fielding at least twenty telephone calls every day asking for assistance.

"It's terrible to be away and it's great to be away," admitted Sally. "It's mind-boggling and it's exciting, it's wonderful and it's awful. But this isn't politics to me. It's a moral issue and a passion. It's a change of life."

The changes that swept through Sally's life surprised her even as she cultivated them. Yet throughout the 11-month battle for Spencerville, Sally had exhibited all the traits of a born organizer—a talent bequeathed upon a mere handful of people like a natural facility for higher math or a terrific singing voice. In abundant quantities, she possessed:

Curiosity

Irreverence

Imagination

A sense of humor

An open mind

And an organized personality, shored up by enough ego strength and "a bit of a blurred vision of a better world," to transform herself into "a well-integrated political schizoid."

This recipe for turning an ordinary citizen into a local leader was first concocted by Saul Alinsky, the godfather of American community organizing. Sally knew Alinsky because Will Collette had urged her to study his organizing techniques and experiences dating back to the 1930s. But Sally could never finish any of Alinsky's books: His writing seemed too theoretical. Sally preferred the real world excitement of backroom strategy sessions and bovine publicity stunts. Yet, despite her aversion to abstraction, Sally concurred avidly with Alinsky's central conviction: American democracy had not failed; in fact, the nation's small towns and poor urban neigborhoods had not yet tested it.

Faith in democracy stood at the heart of the movement for environmental justice in which Sally now found herself immersed. Unlike the European tradition of intellectual leftism, with its penchant for arcane doctrinal disputes and authoritarian rule, the homegrown vision of social change handed down by Alinsky, and picked up throughout toxic America, spawned directly from the ideas of Jefferson, Madison, and Tocqueville—all of whom feared that the United States could lose its unique democracy through citizen inaction. In practical terms, this ideology took root in "people's organizations"— small workshops of democracy that focused on neighborhood issues and flourished under local leadership, but usually required an outside organizer to spark the initial action.

"If a People's Organization were to be thought of as a tree," explained Alinsky, "the indigenous leaders would be the roots and the people themselves the soil. To rest on the soil and be nourished by the soil, the tree must be deeply and well rooted. . . ." An organizer in a people's organization must be an "abrasive agent to rub raw the resentments of the people of the community; to fan the latent hostilities of many people to a point of overt expression; to search out the controversial issues, rather than avoid them. . . ."

Although profoundly influenced by Alinsky's methods, the toxics movement did not swallow whole his ethos and attitude. Alinsky had served an extraordinary political apprenticeship, studying Chicago's gangland underworld—even courting Frank Nitti, Al Capone's "enforcer" and de facto head of the mob after Capone departed for Alcatraz in 1931—in order to cull insights into group behavior. Frank Nitti "took me under his wing," bragged Alinsky. "I called him the Professor and I became his student."

Given this lineage, the Alinsky method of grassroots organizing unsurprisingly attracted more than its share of strident, aggressive, difficult characters. "A People's Organization is a conflict group," Alinsky declared unapologetically. "This must be openly and fully recognized. Its sole reason for coming into being is to wage war against all evils which cause suffering and unhappiness.''

But the women who emerged as the grassroots environmental movement's leaders rejected the brawling, macho style that had evolved over five decades of labor, civil rights, and anti-war organizing. Women in the toxics movement dressed up Alinsky's real politik with compassion and grace so that it could travel more comfortably thoughout small-town America. In the best possible sense, they feminized his methods, making politics more personal and accessible.

To an equal extent, the women leaders of the toxics movement also drew upon a powerful, vastly underrated, and distinctly *female* tradition of American social reform. Since the Revolution, American women had organized intricate, if largely uncelebrated networks of voluntary associations that addressed all variety of human needs—and occasionally tackled the nation's most vexing social problems. By the 1780s, women had formed countless small church committees to provide charitable relief to poor families. In the nineteenth century, women set up groups to "save prostitutes," with more than four hundred chapters of the American Female Moral Reform Society established nationwide. Women pushed forward the struggle for abolition, founding a Female Anti-Slavery Society in Philadelphia in 1833. Two weeks after the Civil War began, women in both the North and South pulled together twenty thousand local voluntary organizations to provide troops with food, clothing, medicine, and money, raising over $15 million worth of supplies during the course of the war in the North alone.

Experience with these causes gave women an opportunity to learn the essential organizational skills of recruiting members, raising money, forming governing structures, and soliciting publicity. "It also sustained and gave political power to certain moral values like compassion and fairness that were eroding in the dominant political culture," observed historian Sara Evans. By launching their battles from voluntary associations located *between* the public world of politics and work and the private world of the family, "women made possible a new vision of active citizenship" that redefined their own identities and simultaneously challenged the larger power structures.

Of all the reform movements driven by American women, the one closest in style and content to the contemporary housewives' campaign of toxic America was the crusade for temperance—a national drive against "liquid evil" and "poison in a bottle." Like the toxics movement, temperance was inspired by health and safety issues. Supporters sought to protect women and children from poverty, desertion, and violence perpetrated by husbands and fathers too frequently disposed to drink. The typical temperance crusader was a married woman in her thirties or forties, living with her husband and children. "In the temperance movement women could view themselves as protecting the home," noted one historian of the era. "In this sense it was a maternal struggle. If they committed public acts, took public stands, it was only in their role as nurturant mothers who must insure a good environment for those dependent on them."

Mary Livermore, a prominent temperance leader, proclaimed that the movement unified women, girding up their moral courage and teaching them "to work intelligently, wisely adapting means to ends . . . (making them) patient, tolerant, able to agree or disagree on minor matters, and yet standing as one where principle is concerned." Frances Willard, founder of the Woman's Christian Temperance Union (WCTU) in 1874, underscored the movement's peculiar female virtues, contrasting her organization's national meetings "with any held by men: Its manner is not that of the street, the court, the mart, or office; it is the manner of the home."

Although deeply concerned with cultivating political power, the Woman's Christian Temperance Union prided itself on maintaining its own version of the "caringness" described today by women in the toxics movement. When a novice organizer botched her maiden speech at the Mississippi state convention, the national Woman's Christian Temperance Union president strode across the platform, embraced her, and welcomed her to the sisterhood. "Men take one line," Willard later explained, "and travel onwards to success; with them discursiveness is at a discount. But women in the home must be mistresses, as well as maids of all work; they have learned well the lesson of unity in diversity; hence by inheritance and by environment, women are varied in their methods; they are born to be 'branchers-out.'"

Like the toxics movement, the campaign for temperance operated on a financial shoestring. The national WCTU held less than $500 in the bank and receipts for one year totaled only slightly over $1,200. Occasionally, roving professional organizers—the Sally Teets of their day—serviced the autonomous, local chapters, offering technical advice drawn from their experience in other parts of the country. But the movement's strength remained its commitment to informal association, as women joined hands first across the streets of their own small towns—and then across the country. Willard assured her colleagues that their movement "was like the fires we used to kindle on the western prairies, a match and a wisp of grass were all that was needed, and behold the spectacle of a prairie on fire sweeping across the landscape, swift as a thousand untrained steeds and no more to be captured than a hurricane.''

Of course, much of this activity remained invisible until it exploded into gaudy events, such as women storming the saloons with hatchets. In the same way, most of the action in small-town toxic America has passed unnoticed by the national news media. Seldom adept at reporting signal events within working-class communities, the national press (when it showed any interest) fixated upon the familiar story of one courageous (or alternately, hysterical) housewife tackling the looming corporate giant—a variant of the David-versus-Goliath chestnut of generic journalism. This story was not so much wrong as incomplete. By focusing on the exploits of a single angry housewife battling a lone dump or incinerator in some obscure part of the country, the press overlooked the political, environmental, and public health significance of a vastly larger phenomenon. By 1992, Citizen's Clearinghouse for Hazardous Wastes would be working with people in seven thousand dues-paying community groups throughout the nation. For many of these women, and fewer men, their unheralded exploits would profoundly alter their personal and political lives.

To Sally Teets, now traveling to toxic sites across the midwest and linked by Citizen's Clearinghouse to contaminated communities throughout the rest of the nation, the shallow efforts of the press seemed bizarre, if not conspiratorial. It was clear from Sally's point of view that the toxics issue was striking the country from every direction. Yet her own experience somewhat explained the detachment and confusion of the media. Before Waste Management threatened to invade Spencerville, Sally never thought about chemicals or toxic wastes. But once the fight began, the subject seemed to confront her at every turn.

When Sally flicked on the television to relax after an evening's strategy session, she discovered the characters on ''thirtysomething'' battling a toxic waste incinerator slated for their neighborhood. At the movies, when she and her husband settled down to watch the Chevy Chase comedy, *Fletch Lives,* they found him coping with the unexpected inheritance of a Louisiana bayou mansion built on top of a toxic waste dump. Not even the Saturday morning cartoon shows eluded the toxic taint: Teenage Mutant Ninja Turtles had burst upon the American cultural scene, mesmerizing millions of American children with the adventures of once-ordinary amphibians whose exposure to a chemical mutagen fantastically increased their size and strength, turning them into Jedhi knights of the municipal sewer system. Teenagers had their own version of the Ninja Turtles in the form of the cult Toxic Avenger movies—a weird saga revolving around a dweebish geek janitor who plunges into a vat of chemical wastes and emerges with superpowers and a three-film grudge against corporate polluters. Produced by the makers of *Surf Nazis Must Die, Rabid Grannies,* and *A Nymphoid Barbarian in Dinosaur Hell,* the Toxic Avenger series took place in Tromaville, New Jersey, the self-declared Toxic Waste Capital of the World.

In truth, the idea of toxic danger had seeped deeper into the national imagination than most people in contaminated communities realized, affecting both popular culture and the higher realms of art and literature. In *White Noise,* novelist Don DeLillo conjured up "an airborne toxic event" that sweeps into a sedate college community as the visible incarnation of the growing dissonance in American life. In Jonathan Franzen's novel, *Strong Motion,* the seismic tremors that threaten Boston are triggered by toxic wastes buried thousands of feet below ground by an unscrupulous chemical company that once supplied defoliants and napalm to Vietnam. In the theater, playwright Steve Tesich, whose earlier film, *Breaking Away,* celebrated the virtues of small-town American life, now drew together buried secrets, discarded lives, and municipal garbage to indict the nation's most poisonous faults in *The Speed of Darkness.* Writers of science fiction and the supernatural also ruminated over the toxic threat. Ursula Le Guin's novel, *Always Coming Home,* concerned a tribe of gentle people called the Kesh who try to live harmoniously with nature in a futuristic, chemically contaminated America. In *Floating Dragon,* Peter Straub updated the ancient curse haunting a creepy New England town by passing an industrial cloud across its skyline that caused people's skin to erupt with chemical chloracne until their faces finally imploded like decomposing pumpkins six weeks after Halloween.

Throughout the 1980s, the image of toxic hazard was expropriated to reflect social problems that had nothing to do with chemicals, waste, or technology. Toxics-as-metaphors now alluded to anything buried and dangerous, perils not acknowledged, dark secrets. A bestselling self-help book exposed the threat of "toxic parents." The phrase "toxic time bomb" was bandied about on television talk shows, referring to any particularly nasty, latent calamity. When the Sunday *New York Times* editorialized about "The Horror Files: How to Manage Europe's Toxic Legacy," the subject wasn't the abundant chemical pollution of the former-communist states, but rather their secret dossiers on private citizens sustained by vast networks of underground police informants.

But while chemical blight might serve as a powerful, if over-adaptable metaphor, the people who actually lived in contaminated towns could not help noticing how often the movies, television, novels, and magazines got the story of *their* toxic encounter wrong.

In the Academy-Award-winning film, *Tootsie,* Dustin Hoffman impersonated a female soap opera actress in order to raise $8,000 and produce his roommate's new play, *Return to the Love Canal*—a preposterous and hilarious premise when the film opened in 1982.

"Nobody's going to do that play," objects Hoffman's agent early in the movie.

"Why?"

"Nobody wants to pay $20 to see a couple living next door to chemical wastes. They can see that in New Jersey."

But by 1990, people *were* returning to Love Canal—now rechristened Black Creek Village—as the government assured prospective homeowners that the country's most notorious contaminated community had been sufficiently cleaned up and its real estate prices suitably discounted.

The strangest aspects of the toxics crisis were blending together, often making it difficult for the most astute observers to distinguish fact from fancy. The disquieting images of toxic America were spreading almost as quickly as the contamination itself.

- - -

When Sally Teets showed up at a contaminated community as a Citizen's Clearinghouse organizer, she usually began her work by meeting with a small group of disgruntled citizens to relate her own story. Invariably, Sally's opening talk included a stern admonition.

"If there's somebody in this room that doesn't believe in their heart that they're on the right side of the issue," she would tell them, "I want you to get up and leave because I don't even want to deal with you. You have to believe in your heart that what you're doing is right."

In the midst of her speech, Sally's own conviction often overflowed into tears. Afterwards, people rose from the folding chairs lined up along the walls of their church basement, or they abandoned the squishy couch in their neighbor's cluttered living room, and they clustered around her. "Even if they don't say a word to me, they'll come up and touch my arm, or they'll pat me on the back, or they'll just brush my cheek with their hand. They want to touch somebody who won."

Sally's talk always culminated with Spencerville's splashy and unequivocal triumph over Waste Management. But she had other stories in her repertoire. Immediately after the Spencerville victory, members of the Mennonite community in neighboring Bluffton asked her to help organize against a proposed "McStop"—a sprawling roadside attraction composed of a truck wash, hotel, and gas station built around a new McDonald's franchise. The Mennonites feared the McStop would attract massage parlours and adult bookstores, as a similar site had seven miles away. "We beat them," bragged Sally. "We beat McDonald's." Later, in the tradition of the temperance movement, Sally "branched out" even further, joining a Citizen's Clearinghouse contingent at the national mobilization against homelessness in Washington, D.C., and marching under the banner of "Safe and Affordable Housing." When the mobilization's leader, Mitch Snyder, committed suicide soon afterwards, Sally attributed his death to "the terrible let-down" he must have felt after the march. "It's tough every time you do a really good gig and it's over," she explained. "A part of *me* dies each time."

But Sally's new status as the roving Midwest Organizer for Citizen's Clearinghouse meant the good gigs would keep rolling in for years. Even when she returned home, the rush of the road did not subside. Organizing animated Sally's speech and gestures, quickened her pulse, transformed her life. Her new job was thrilling, risky, unpredictable, sexy. Beyond all that, it just made plain sense.

"If you're fighting Waste Management," she explained, "and I've already fought them, there's no reason why you should have to reinvent the wheel. I'll tell you the tactics that worked for us, you tell me your best tactics. I'm not a knight on a white horse who rides into town and fights the people's fight for them. I don't say, 'You've got to do this and you've got to do that.' All I say is, 'This is what worked for me. Take what's appropriate for your fight and use it. If it doesn't work, throw it in the trash can.' "

Roaming across toxic America significantly increased Sally's own stock of tactics. In Michigan, she heard about a dozen elderly women who protested against being charged for a new household water system that would bypass the contaminated municipal source. Dressed in bathrobes and curlers, the women appeared at city hall one morning, demanding to use the officials' bathrooms to shower and conduct their morning ablutions. In Appalachia, she talked to citizens who registered their disapproval of the local politicians' decision to keep dumping chemicals into their waterways by shoveling steaming manure onto the city hall steps. People in New Jersey told Sally how they kicked off an official meeting by singing "America, the Beautiful" with the lyrics amended to provide public testimony that government officials had refused to hear. Women in California told her how they had stopped the state's attorney general from harassing their organization by going door-to-door around his suburban block, informing neighbors about his attack on "ordinary citizens"; then they picketed his front gate all afternoon while he tried to show his house to prospective buyers.

The cross-pollination of strategy and tactics pointed to the activists' greatest strength. As people in contaminated communities reached out to one another, their sense of isolation dissolved. Individuals regarded themselves less as victims than as part of a movement; suddenly, they were toxic crusaders.

Government and industry critics took an entirely different view, condemning the best-organized communities as the worst examples of NIMBYism—the Not In My Back Yard syndrome. They labeled these communities selfish, short-sighted, socially irresponsible, even psychopathic.

"The NIMBY syndrome is a public health problem of the first order," declared the Southern California Waste Management Forum. "It is a recurring mental illness which continues to infect the public. . . . Organizations which intensify this illness are like the viruses and bacteria which have, over the centuries, caused epidemics such as the plague."

But direct contact with the obstructionist communities revealed a more complicated story. According to journalist Charles Piller, who studied a range of NIMBY groups intent on blocking everything from nuclear weapons production to biotechnology, the supposed syndrome was "merely a symptom. It has grown out of a scientific and technological enterprise that tries to function like a world apart from the people it affects, spinning out risks and benefits without attention to their fair and equitable allocation. . . . Without trust, people withdraw consent from those who run society. NIMBYism demonstrates a gradual withdrawal of consent at the grass-roots level."

The NIMBY groups were turning around the equations of risk-benefit analysis proffered by government and industry statisticians, vigorously pointing out that the people subjected to most of the risks seldom enjoyed an equal share of the benefits. *We are the canaries in the coal mine*, the NIMBY groups declared, employing a metaphor that quickly spread through contaminated communities everywhere. Like the legendary canaries whose asphyxiation in the coal mines indictated danger to the workmen, many residents of small-town toxic America believed they were the harbingers of a health crisis that would soon plague the entire country.

Moreover, the NIMBY characterization only made sense when communities were observed in isolation. As NIMBY turned into a mass phenomenon—into a movement—its essential character, tenor, and demands decisively altered. Over the past decade, communities that once cried NIMBY have altered their position to NIABY and NOPE; that is, the demand of Not In My Back Yard has been extended to Not In Anybody's Back Yard or Not On Planet Earth. "I want people to say, 'God, this is happening everywhere,'" emphasized Sally Teets, "'and we're going to have to do more than just save our own butts.' Change will come from the bottom up. Local issues will change state issues. State issues will change federal issues."

When pressed for details, most toxics organizers drift into comfortably vague pronouncements. "If the people want change," declared Sally Teets, "it will come." But the key to their national strategy, such as it exists, is the notion of "plugging up the toilet"— blocking toxic waste disposal sites and incinerators throughout the entire nation, thus forcing some kind of fundamental reckoning with society's toxic overload. On the way to this goal, people in thousands of small towns across the country are learning to work together.

It was our increasing awareness of this journey from local intransigence to national resistance that originally drew us to Ohio.

Before leaving California, we had talked by telephone with Patty Wallace in Lawrence County, Kentucky, about her own progress as a grassroots toxics organizer. In 1983, Wallace heard about a "recycling plant" planned by a company named Pyrochem for the nearby town of Louisa. Wondering why Louisa, with a population of two thousand people needed a recycling plant, Wallace began to ask questions. She learned that the proposed plant was headed by Jim Neel, a man of some fame in Kentucky. As the director of Kentucky's Atomic Energy Authority in the 1960s, Neel originated the idea of rescuing the state's economy by offering it up as a "service station" for radioactive wastes. Neel later became director of the private company operating Maxey Flats, Kentucky's notoriously contaminated radioactive waste disposal site.

Eventually, Wallace learned that the "recycling plant" slated for her community was actually Pyrochem's term for a hazardous waste incinerator. After burning the hazardous wastes, Pyrochem wanted to "recycle" the contaminated incinerator ash to build roads throughout Kentucky.

For five years, Wallace and 75 members of her local organization fought Pyrochem. In 1988, they finally persuaded the Kentucky legislature to write a new law changing the jurisdiction for hazardous waste incinerator permits from state to county control. With the new law in hand, Lawrence County promptly tossed Pyrochem out of Louisa. If the story ended there, the NIMBY—Not In My Back Yard— label might well apply to Wallace and her friends.

But shortly after her own community's victory, Patty Wallace heard on TV that Mason County, West Virginia, would soon get its own "recycling plant"—courtesy of Pyrochem. "My ears perked up," Wallace recalled. "Once you learn about stuff like this, you can't just forget it." Two days later, Wallace hit the road for West Virginia.

In Mason County, she met Nancy Anderson, a housewife who had her own doubts about the recycling plant. Pyrochem had promised to collect and dispose of household toxics such as paint, solvents, and bleach, billing itself as a community service. After talking with Wallace, Anderson launched a petition drive against the incinerator. "Patty Wallace and her group had been fighting Pyrochem for five years," explained Anderson. "I don't think they wanted anybody else to go through the same thing." Anderson collected six thousand signatures in 30 days. Pyrochem departed Mason County, West Virginia.

Like Patty Wallace, Nancy Anderson found that the local skirmish whetted her taste for regional battles. When she heard that a small farming community in Nova, Ohio, had also taken up against a toxic waste incinerator, she volunteered her services. On July 2, 1989, Wallace traveled to the rural township of five hundred residents. When she got there, she found local organizers like herself from small towns in Ohio, Michigan, Indiana, Illinois, Virginia, West Virginia, Kentucky, Pennsylvania, and Tennessee. Nova residents had drawn together people from nine states to brainstorm a strategy for keeping the toxic waste incinerator out of rural Ohio. What they came up with were ideas that could be used to fight incinerators slated for communities across the country.

When we arrived in Nova one year later, the visitors had long departed, but evidence of the fight remained everywhere. At nearly every house, we saw bright red signs posted in the front yards, reading "Stop IT"—the IT being International Technologies, the company that had first proposed the incinerator for Nova. Almost every person in town had united behind the local citizens' group opposing the incinerator. In the most recent election, the county trustee who had stubbornly backed the incinerator found himself booted out of office by a 78 percent margin, and replaced by an anti-incinerator candidate.

At the crossroads dividing the township, we met with nine Nova citizens, ranging in age from 29 to 67, in the small trailer that served as headquarters for the Stop IT campaign. Most of the Stop IT veterans had visited Spencerville during its fight against Waste Management. In turn, Sally Teets had traveled to Nova several times, working with the group on its initial strategy.

Nova, Ohio.

The folks from Nova also informed us that everybody in town knew about our arrival. Since the fight had begun, Nova regarded uninvited strangers with suspicion—if they didn't carry scars from the small-town toxic battleground. Residents had grown sadder, wiser, angrier, and immeasureably less certain about the future. Soon everybody was reminiscing about the less stressful life they had led before the incinerator's threat.

"Those were fun times!" exclaimed Diana Schlaufman, one of Stop IT's leaders. Her neighbor, Elaine Drotleff, enthusiastically agreed.

Yet each member of Stop IT also thought that the battle against International Technologies had given their commmunity something of unexpected value. The ordinary obsessions of small-town life—work, family, church affairs, and high school sports—hadn't drawn people together in the same vigorous, intimate way as the outside threat of the incinerator. In the fight, Nova rediscovered itself.

What was it like working together over the past two years? we asked the nine citizens in the trailer.

Lois Kinter, a 42-year-old housewife and mother, considered the question.

"This is . . . ."

"War!" shouted Scott Medwid, the group's youngest member.

"No," insisted Lois, "we're like a family. We get together and have our problems, but we continue on. It's like a good marriage. There's a close-knittedness here. And I know that if I died tomorrow, my family would say, 'Well, maybe she didn't always have supper ready and the socks paired, but at least she did what she could to help humanity.' "

"Our lives have changed," agreed Sam Lyle-Medwid, Scott's wife. "And when we finally win, it won't be like taking off a coat and saying, 'Okay, I'll hang it up now and go on.' Our lives will never be the same."

We asked if anybody felt angry about the hundreds of hours taken from family and leisure time.

"We got in this situation," explained Franklin Rickett, a local farmer, "because people like me laid back, minded their own business, and didn't get involved in government. That's our own fault. But we're going to change that now."

How far will you go? we wondered.

"I don't think we know," answered Lois.

"I know," said Scott. "I'm not going to take up arms. I'm not going to kill anybody over this. It's not going to come to shooting."

If they started to build the incinerator would you sit down in front of the bulldozers?

"*Yes,*" they answered in a chorus.

Would you go to jail?

All, again: "*Yes.*"

"As a group, we always intended to work within the process and do things lawfully," insisted Diana Schlaufman. "But if it comes to the point where the bulldozers arrive on the site, I'll be in front of the bulldozers. If I'm arrested, it won't be because I've done anything wrong. I know I have a right to protect my life and my family and everyone else in this community."

But didn't anybody think the government would protect them?

The citizens of Nova stared at us as though we were insane.

"There is a conspiracy between government and industry," declared Vern Hurst, a 27-year retired career officer in the U.S. Air Force.

"As citizens, we're as guilty as anybody else," insisted Diana. "We sat back with that apathetic nature that citizens across the country seem to have, saying to ourselves, 'The EPA is there and they'll solve all the problems.' We let the EPA become what it is today. We didn't do our part. We've allowed our rights to be taken away from us without even realizing they're being taken away. It's time to take them back."

"I never thought I'd be fighting my government," admitted Elaine. "That's basically what I'm doing. I'm fighting my government."

- - -

Some of the harshest words spoken today about the American government can be heard in the nation's small towns, once bastions of love-it-or-leave-it patriotism. In many of these towns, this disenchantment with government can be explained, at least partially, by the desultory history of the EPA, an agency uniquely adept at converting apolitical housewives and farmers into uncompromising radicals overnight.

Founded by President Nixon on the heels of Earth Day in 1970, the EPA drew together a welter of powers formerly held by the departments of Interior, Agriculture, Health, Education, and Welfare, and several executive bodies. As an offshoot of Nixon's New Environmental Policy—regarded by many as a feint against the unpopular Vietnam War—the EPA was charged with aggressively monitoring and reducing all varieties of environmental hazards. Conceptually, the invigorated federal role in protecting the nation's skies, lands, and waters couldn't have been more popular at the time. "Ecology has become the political substitute for the word 'mother,' " observed Jesse "Big Daddy" Unruh, the California legislature's ultimate insider. But the EPA's role in American economic life remained murky, particularly when it came to regulating industry and agriculture.

Perhaps a clearer mandate would have evolved over time had the business of government proceeded as usual. But as the second Nixon administration descended into the chaos of Watergate, the executive branch wholly ceded its responsibiity for watchdogging the environment to Congress. With more enthusiasm than understanding, both houses approved during the 1970s a small mountain of pioneering environmental legislation, not quite grasping its cost or potential for conflict. By the time Jimmy Carter took office as the first self-proclaimed "Environmental President" since Teddy Roosevelt, the competing strands of legislative activism had tightened into a Gordian knot of bureaucratic complexity. The new executive displayed little enthusiasm for slicing through its tangle of laws, lawyers, and loopholes.

The Toxic Substances Control Act, for example, was enacted to protect Americans from dangerous chemicals utilized by industry and agriculture. But the legislation also presumed that tens of thousands of chemicals already on the market were safe, unless proven otherwise by the EPA. Unfortunately, the agency had neither the time nor money to test scores of thousands of chemicals already in common use, never mind the one thousand to two thousand new chemical products introduced into the marketplace every year. Instead of increasing the EPA's budget—and therefore, its power—Congress instructed industry to conduct its own tests, quantifying the health risks of their new products, and passing their findings along to the government. Unsurprisingly, this approach provided extraordinary opportunities for abuse. In 1977, federal investigators revealed that the nation's largest commercial testing firm, Industrial Bio-Tech Laboratories, Inc., had rigged the data in hundreds of government-mandated tests to assess the safety of potentially hazardous chemicals in order to give their clients a clean bill.

The EPA further eroded its credibility by undertaking a "toxic of the month" research method. Individual chemicals seemed to attract the agency's notice only after government outsiders—not EPA officials—publicly indicted their risks to human health. In this mannner, the EPA tackled dioxin, PCBs, and various pesticides, but always with the appearance of playing catch-up to citizen outrage.

If the Carter administration reluctantly implemented the complex and expensive laws meant to guard the public from toxic harm, the Reagan administration robustly set about demolishing them. "Government is not the solution to our problem," the new president declared in his first inaugural address. "Government is the problem." Accordingly, the Reagan administration softened the bite of environmental regulators by extracting crucial staff and funding as though they were impacted wisdom teeth.

To a large degree, the Reagan victory reflected the American public's own disengagement from civic life. Yet the new adminstration's anti-environment antics rattled many people—particularly, those citizens living in the toxic towns now turning up throughout the nation. Ironically, the radical retreat from federal protection of the environment during the 1980s coincided with the discovery of a dark wealth of chemically-contaminated communities, from dioxin-coated Times Beach, Missouri, to the mysterious cancer clusters of Woburn, Massachusetts, and McFarland, California. Throughout the 1980s, thousands of grassroots anti-toxics groups sprung up throughout the country.

Yet Reagan's new EPA administrator, Anne Gorsuch Burford, evinced no interest in these alarms. Instead, she set about dismantling the EPA, particularly blunting the effects of toxics legislation. With EPA funding initially cut 12 percent, Burford administered the shrinking agency with the budgetary ruthlessness of a provincial sovereign supressing an occupied nation. Not surprisingly, the agency's most experienced staffers defected; the annual attrition rate ran 32 percent. "It is hard to imagine any business manager consciously undertaking such a personnel policy unless its purpose was to destroy the enterprise," declared Russell Train, the EPA's former administrator under Nixon and Ford, in a *Washington Post* op-ed piece. "Predictably the result at EPA has been and will continue to be demoralization and institutional paralysis."

Before heading off to prison for perjury, Rita Lavelle, the EPA's assistant administrator for Solid Waste and Emergency Response, unwittingly detailed the true direction of the agency in her own written statement of personal goals: "1. Change perception (local and national) of Love Canal from dangerous to benign. 2. Obtain credible data that (1)50 of nation's most dangerous hazardous waste sites have been rendered benign. 3. Provide credible proof that industries operating today are not dangerous to the public health." Prior to the EPA, Lavelle had served as a public relations flack for Southern California's Cordova Chemical Company. Her view that the nation's environmental problems could be resolved through the magic of public relations reflected the perverse fantasy life that separated Reagan administration deregulators from the growing number of ordinary citizens who now feared for the health of their communities.

Yet all these problems paled in comparison to the disaster of Superfund—the jewel in the crown of EPA mismanagement that continued into the 1990s to disappoint and infuriate almost everybody involved with toxic wastes.

Superfund had been originally promoted as the quick-hit, short-term solution for cleaning up the nation's most dangerous toxic waste dumps, leaking landfills, and blighted industrial neighborhoods. First conceived as a limited five-year project costing $5 billion, the program relied on a national trust fund garnered from taxes paid largely by the chemical and petroleum industries. This "super fund" would underwrite the price tag of clean up when damages could not be recovered from the contaminated site's original polluters. From the beginning, Superfund proved a classic case of best intentions run aground.

Following Reagan's 43-state landslide victory in 1980, President Carter's lame duck administration could not muster the clout to push the program's original conception through Congress. As a result, Superfund wandered around Washington as a political orphan for several crucial weeks. Finally, the outgoing president signed into law a compromise bill slashing Superfund's budget from $4.1 billion to $1.6 billion—thus guaranteeing that even the most modest estimations of the nation's toxic overload would not be quickly dispatched. Other flaws inherent in the program's original design further stalled action. Superfund's sponsors had lauded the principle of "Shovels first, lawyers later"—a promise to evade immobilizing legal battles by immediately beginning the engineering work needed to reclaim contaminated sites. But legislative provisions making any one polluter responsible for the full costs of redeeming a Superfund site—even if the polluter had not violated any laws—made inevitable the storming of Washington by litigating hordes.

To make matters worse, nobody really knew—or knows—how many sites required immediate attention. Shortly after Love Canal, aerial photographs taken by the EPA pointed to one hundred locations on Staten Island alone that appeared to be unmarked waste sites. "Abandoned wastes were uncovered seemingly everywhere beginning in the late 1970s and early 1980s," admitted a high-ranking EPA scientist. But during the Reagan years, top Superfund staff told Senate investigators that EPA chief Burford had strained to give the impression that toxic wastes were not an urgent problem. She even wanted to leave money in the coffers—an unthinkable act in federal agencies—thus diminishing the chances of Congress renewing the program.

Today Superfund lists over twelve hundred priority projects. Yet this number indicates only a fraction of the sites anticipated to become future disasters. In 1985, the Congressional Office of Technology Assessment warned that ten thousand contaminated sites might eventually require clean up. By February 1991, after 13 years of operation, Superfund had only managed to fully redeem 33 sites. At this rate, the initial twelve hundred sites would be cleaned up in about five hundred years.

To judge from the government's own estimations, not only the quantity but the quality of Superfund work also lagged far behind expectations. According to one of four devastating critiques conducted over the past decade by the Congressional Office of Technology Assessment, Superfund is "a loose assembly of disparate working parts; it is a system of divided responsibilities and dispersed operations. There is no assurance of consistently high quality studies, decisions, and field work or of active information transfer."

"This is a program that hardly ever gets anything right," summarized Joel Hirschhorn, an environmental consultant who headed up a blistering report to Congress in 1988.

Other studies have indicted Superfund efficiency, questioning whether the program protects human health and the environment. "Nowhere did I encounter a passionate determination among the contractor specialists, in particular, to clean up the wastes as rapidly as possible. . . ." admitted Glenn Schweitzer, former director of the EPA's Office of Toxic Substances.

People overseeing cleanup programs invariably blame the "system"—the baroque configuration of state environmental agencies, regional EPA offices, and the federal bureaucracy. At times, the chief duty of these government branches seems to be debating who will pay for the costly clean ups. Some analysts now believe Superfund costs may exceed the savings and loan debacle, running to $1 trillion over the next 50 years. And nearly one quarter of that amount, about $200 billion, will be spent on "transactional costs"—meaning, for the most part, lawyers' fees. No longer supportable through the original scheme of fines and industry taxes, the inflated cleanup bill will probably require a taxpayer bail-out.

If all this wasn't enough to stoke the bonfires of cynicism about government, citizens in contaminated communities also observed at ground level the revolving door that spins regulatory staffers out of the EPA and into employment in the toxic waste industry. The most formidable example is former EPA chief William Ruckleshaus, who now serves as the president of Browning-Ferris Industries—the nation's second-largest hazardous waste handler, until getting out of the business at the beginning of the 1990s. But equally corrosive to trust have been the lower-level EPA and state regulatory staffers who use their government experience as a step up into far more lucrative industry positions.

"A lot of regulatory people switch over," complained Linda Young, whose town of Toone, Tennessee, was contamined by illegal industrial dumping. "So they aren't going to be too hard on the industry while they're working with the government. You think these officials are here to help you—and the next thing you know is they're working for the company that did the damage."

▪   ▪   ▪

Given their alienation from government, it would seem natural that people in Louisa, Kentucky; Mason County, West Virginia; Nova and Spencerville, Ohio—and all the other towns of toxic America— would align themselves with the nation's major environmental groups. Yet for the most part, the environmental boom of the 1980s bypassed America's small towns and rural communities—the places most likely to contain a hazardous waste incinerator or leaking toxic landfill.

Spurred on by the Reagan administration's hostility to environmental regulation, membership in national environmental organiza-

Toone, Tennesse, where hazardous wastes contaminated the water supply. Tom and Linda Young, with their daughters Laurie and Kelli. "You think pollution is cities, smokestacks," said the Youngs. "In New York, people are afraid to go out of their houses. Well, I'm afraid of what's come into mine."

tions swelled from four million to seven million during the 1980s. Some established groups, such as the Environmental Defense Fund and World Wildlife Fund, doubled in size. Yet as the mainstream organizations expanded, their leadership decisively shifted from cadres of outraged activists to a rising class of professional managers, publicists, fund-raisers, and lobbyists. No longer regarded as a kooky cause, environmentalism now offered a career ladder into a comfortable, middle-class life-style. Lawyers assumed the top administrative positions and salaries rose accordingly. By 1992, the president of the National Wildlife Federation earned $250,000—more than 15 times Sally Teets' annual salary.

This slippage from a fractious protest movement into a stable political institution signaled an estrangement from the front lines of toxic America. In an attempt to consolidate power, the large environmental groups addressed most problems in a single voice—and that voice distinctly lacked the rough cadence and colloquialisms of the working-class towns hit hardest by chemical exposures.

"The Sierra Club is becoming like Velveeta," complained David Brower, formerly the organization's director, "everything must be processed." The National Resources Defense Council, National Audubon Society, and National Wildlife Federation erected new head-quarters in New York and Washington, D.C., drawing fire from the grassroots over the sway of "Potomac conservationists." Environmental respectability also smoothed the way for the leaders of large non-profit organizations to identify with their peers in the profit-making sector—including executives in corporations viewed by many people in toxic towns as environmental enemies. The chairman of Waste Management, Inc., whose hazardous waste sites received fines for violations running into the millions (and whose dump Sally Teets repelled in Spencerville), joined the board of the National Wildlife Federation and became a prominent donor. Television documentaries produced by the National Audubon Society secured funding from General Electric, the company responsible for the greatest number of Superfund sites. Meanwhile, groups like the Dumpbusters relied on weekend cow crap-outs for funds.

By the end of the decade, many people in contaminated communities viewed the leaders of national environmental organizations as closer kin to government bureaucrats than toxic crusaders. But small-town toxic America had another reason to feel alienated from mainstream environmentalism. People living in contaminated communities vigorously rejected the strain of misanthropy that had run through the environmental movement since its infancy.

"Pollution, defilement, squalor are words that never would have been created had man lived conformably to Nature," wrote John Muir, in an early articulation of environmentalism's fundamental skepticism about humanity. "Birds, insects, bears die as cleanly and are disposed of as beautifully. . . . The woods are full of dead and dying trees, yet needed for their beauty to complete the beauty of the living. . . . How beautiful is all Death!"

People in contaminated communities could not rationalize the death of their children as an ecological event. Concern for their families had drawn them into the fight. If mainstream environmentalists insisted that *people* were the real problem—since they created most pollution—then people like Sally Teets and the citizens of Nova, Ohio, would take it upon themselves to redirect the conversation. The source of environmental degradation, asserted this new breed of grassroots activist, remained the ordinary citizen's inability to affect the critical political decisions shaping his or her life. At bottom, the issue was power.

For most mainstream environmental groups, these larger questions of democracy held only limited interest. When Nova residents first took on International Technologies, they dispatched 150 letters, seeking help from all the well-known environmental organizations. Most responded with fund-raising brochures or public relations information about their own agendas. Many didn't respond at all. Only Greenpeace and Citizen's Clearinghouse for Hazardous Wastes grasped Nova's dilemma and immediately offered assistance. The critical assistance came from people much like themselves, already scarred and transformed by similar battles.

Yet the detachment with which mainstream environmentalism regarded toxic America made sense, given the movement's genealogy. American environmentalism has always been an elite affair, inspired and directed by society's most privileged ranks.

When Teddy Roosevelt recruited a select number of "American hunting riflemen" in 1888 to form the Boone and Crocket Club—whose Congressional clout soon secured the nation's first big game refuge in Yellowstone—the exclusive character of the "conservation" movement had already crystallized.

One hundred years later, environmentalists drawn from society's more affluent ranks would still dominate the scene, regarding with suspicion the attitudes and occupations of working-class America. After all, wasn't the blue-collar American the person who chopped down the forests, hunted the wildlife, farmed the pesticide-laden soil, labored in the factories that fouled the air, filled their gas-guzzling Detroit behemoths with more kids than the world could sustain—and didn't give a damn about the exhortations of the more enlightened, better educated brethren?

The siting of toxic waste dumps and incinerators underscored the subtle divisions of class in America. Most citizens preferred not to think about class—or if they did, they invariably insisted that they belonged to the huge, amorphous, and incessantly celebrated middle-class. But over time, the residents of contaminated communities were forced to reexamine their place in American society.

Not only had their government abandoned them. Not only had the large environmental organizations—Sierra Club, Friends of the Earth, the Audubon Society, and the rest—shown no interest in their resistance. Not only had the liberal, middle-class, media establishment snobbishly caricatured, ignored, and rejected what Christopher Lasch calls "the positive features of petty-bourgeois culture: its moral realism, its understanding that everything has its price, its respect for limits, its skepticism about progress." The truth about toxic America was considerably worse. People in small towns and working-class communities were actually being targeted for toxic waste dumps and incinerators that would not be politically feasible in more affluent locations.

In the 1980s, the hazardous waste industry commissioned a study by the Los Angeles consulting firm, Cerrell Associates, to identify the places least likely to resist a hazardous waste site. Unsurprisingly, the report suggested that facilities should be placed in areas where residents were "not concerned with issues." According to Cerrell, those Americans most thoroughly disengaged from civic life usually proved to be Republican, blue-collar, Catholic, and low-income. They were above-middle-aged, high school educated or less, and in their political opinions they seemed favorably disposed to a "free-market orientation." Almost always, these perfect victims congregated in small towns or rural areas.

"The real reason a neighborhood gets a toxic waste dump," summarized Will Collette, chief organizer with Citizen's Clearinghouse for Hazardous Wastes, "is because a Los Angeles consulting firm says they are too stupid to resist."

Yet throughout the 1980s and early 1990s, events proved the Cerrell study to be consistently wrong. The people of small-town America who were predicted to roll over turned out to be the industry's most implacable foes—particularly when they joined forces. In truth, it was the more educated, white-collar, suburban middle-class that more often proved effete and inept.

When we visited Uniontown, Ohio, located less than one hundred miles northeast of Nova, we glimpsed from another perspective how class differences ultimately shaped the style and strategy of all political calculations within the anti-toxics movement.

Uniontown was a middle-class bedroom suburb of Akron, a speck of comfortable old homes and tract houses bordering the commuter highway. Compared to Yukon, Nova, or Spencerville, it looked like a rising, prosperous community—a step up the social ladder, requiring some measure of financial success. At the center of Uniontown stood the aptly named Industrial Excess Landfill—750,000 metric tons of toxic waste deposited throughout a 30-acre dump. At one point, the methane venting up from the dump collected inside a suburban home and it exploded. Both the polluters and the U.S. goverment had deemed the site too costly to clean up. In short, the problems of Uniontown seemed sufficiently vast and dramatic to warrant the predictable clamor of a well-organized and outraged citizenry.

We spent an evening with five members of Uniontown's local anti-toxics organization. Unlike the farmers, miners, laborers, and housewives we had encountered in other communities, the Uniontown group exhibited a distinctly professional air. They were employed as engineers, teachers, technicians, small business people; they came to the meeting dressed in sports coats and ties, cashmere sweaters and heels. But most surprising was the committee's conversation, indicating its remarkable disconnection from the people it claimed to represent.

Much of the evening's discussion focused on the intricacies of EPA policies and the machinations of federal regulators.* The five Uniontown residents spoke a strange, acronymous language, indecipherable to most outsiders—gabbling away enthusiastically about CERCLA, SARA, FIFRA, RCRA, TSCA, and other equally arcane allusions to government regulations and bureaucracies.

Another leading topic turned out to be the desperate need to fill in the "data gaps" that would finally establish the validity of the health threat to their community. The group had become an extraordinarily adept research body, penetrating the vagaries of organic chemistry, hydrology, and biostatistics. "We knew we had to gather information," insisted Chris Borello, the group's leader. "We became like sponges." Over time, even the EPA sought them out for information. But what the group seemed not to have learned—in fact, forcefully rejected—was the central lesson of almost every successful community group's long struggle: The battle was not technical, scientific, or medical; it was political.

After eight years of activity, the Uniontown committee appeared to be getting nowhere. "They've just kept stalling us out, stalling us out, stalling us out," admitted Borello.

"People like us don't change the system," insisted another member, arguing that direct confrontation was messy and ineffective.

"What we need is a powerful person at the top who can apply pressure," suggested another, rejecting the notion of grassroots power. "It has to change at the top."

Uniontown seemed a textbook case of a community pointed in the wrong direction. Over the past year, the group had traveled twice to the regional EPA office in Chicago, three times to Washington—everywhere but around the corner of their own town to organize a large, clamorous resistance from the ground up. When one committee member mentioned his declaration to an EPA official that four hundred Uniontown families would have to be bought out by the government, another admitted, "We don't even know if four hundred familes want to be bought out." The dedicated, but seemingly disconnected committee of Uniontown appeared to identify too thoroughly

---

* These acronyms stand for: the Comprehensive Environmental Response, Compensation, and Liability Act (better known as Superfund); Superfund Amendments and Reauthorization Act; Federal Insecticide, Fungicide, and Rodenticide Act; Resource Conservation and Recovery Act; and Toxic Substances Control Act.

with its peers in government and industry to stage a real rebellion. Confrontation and pressure politics were regarded as undignified, *declassé*. The committee kept a strange faith with progress, hoping that technology would eventually solve their problems.

"I can walk into some towns," concurred Sally Teets, "and see they don't have a rat's chance in hell. If they're relying on attorneys, geologists, and technical facts, they're going to lose. That makes everything negotiable. Then when something new arises, the rules of the game change, and they're going to get jerked around. But when I meet with other communities, I can walk away knowing they're going to win. They're always the people who are talking to their neighbors, building power, and telling themselves, 'There's no way we're going to lose.'"

In the process of battling big business and big government, the people of small-town toxic America had immersed themselves, usually for the first time, in the sticky stuff of democracy. Like the transubstantiated hero of the Toxic Avenger films, they emerged ready to fight—and armed with a ferocious new faith in the country's resiliency.

As we traveled throughout contaminated communities, we asked people about their perceptions of America today. Was the country in decline? Was it less democratic than 50, 20, or even 10 years before? Almost every person involved in a strong community organization thought our questions sounded foolish.

Back in Nova, Ohio, when we asked the members of Stop IT if they believed the country had slipped, Scott Medwid got angry.

"I think that's a lot of b.s.," he declared earnestly, glaring at us from his chair in the Stop IT trailer. "You see it all around." He gestured across the trailer at his eight friends, his new family, his troops. "We're getting *back* to what this country was."

"It's Like a Murder Mystery: We've Eliminated the Butler— But Who Did It?"
—McFarland, Spring 1988

Sign on open farmland two blocks from the housing tract in McFarland, California.

On January 19, 1988, the California Department of Health Services released the first draft of its Phase II study of McFarland's cancer cluster. Shouldered by the state because of the public outrage following Kern County's inconclusive report, the new study represented one of the most ambitious efforts undertaken anywhere to identify the source of a childhood cancer cluster.

Once again, throughout McFarland, anticipation over its findings raised hopes that the town's health problems would soon be identified and ultimately eradicated.

The state had conducted a case control study, comparing the family backgrounds, homes, and health histories of 10 McFarland children suffering from cancer with 20 healthy children from the same community. To be included in the study, the cancer cases had to have been diagnosed during the 11-year study period, from January 1975, to December 1985. The study only counted children who lived in McFarland at the time of their diagnoses.

The time and age boundaries established by the state eliminated from the study some children with cancer who the most concerned citizens in McFarland felt should have been included. For these parents, the study appeared flawed from the start. Their own count of kids with cancer included at least five other cases—children who were either too old to qualify for the study, or who had moved from town at the time of their diagnoses.

The Department of Health Services persistently explained that every scientific study needed to set limits *somewhere*, but this argument did not comfort veterans who already regarded the logic of state and local bureaucracies to be, at best, Byzantine.

Conducted between 1986 and 1987, the Phase II study matched the cancer kids and a healthy control group in terms of birth dates, sex, and ethnicity. Of the cancer cases, 10 out of 13 children—about 77 percent—were Hispanic. Since the town's overall Hispanic population stood around 77 percent, this proportion did not strike the researchers as remarkable. Perhaps more significantly, the state officials never brought up the surprising fact that Hispanics in the United States actually have *lower* rates of cancer than the national average— making the McFarland cluster even more significant. The low cancer rate for Hispanics contradicted the homespun theories blaming the children's illnesses on some peculiarity of race or culture.

At the beginning of the study, Department of Health Services researchers interviewed parents (almost always, the mothers) about their children's life histories. They discussed the children's illnesses and medications, where they had lived and attended school, the places where they spent their spare time, diet and consumption of water and other beverages, specific exposures to toxic substances, details about their home's building materials, and information regarding their heating and air-conditioning. The study also questioned parents about their own diet, medical histories, hobbies, and workplace exposures to toxics. Mothers considered the chemicals they might have encountered during pregnancy. Fathers traced their contact with toxics three months prior to conception, when their sperm was forming.

Despite the complexities of the survey, the information on the children offered a far better chance for locating the source of the cancers than a comparable study involving adults. The kids didn't smoke or drink, nor had they spent years zigzagging across the country, laboring in pesticide-laden fields or among industrial workplace contaminants. Given the resources marshalled for this comprehensive effort, many researchers talked enthusiastically about immersing themselves in, and perhaps even solving, the cancer cluster mystery.

To begin, the scientists focused their attention on what they had come to regard as the "window of opportunity"—a period between 1980 and 1983, when the children might have encountered something in the environment, or several things, that caused the cancers. The state team delved into a vast array of environmental investigations, combing the town and its outskirts for evidence of toxic exposures.

Researchers collected soil samples down to a level of three feet at nine cancer case homes and six homes randomly selected as controls. They dug up the dirt in local schools, parks, and roads. Tests indicated trace levels of heavy metals and pesticides, but they did not appear sufficient to pose a health threat.

The state team returned to McFarland's water supply.

In 1987, the pesticide DBCP had been found in one well servicing the town. Subsequent tests indicated a probable lab error, but the well was shut down soon after because of unacceptable levels of sand pumped from its depths. Still, McFarland's residents had always suspected water contamination—particularly given the immense problems throughout the Central Valley, where at least 2,400 wells were polluted with DBCP. The researchers collected water samples from six municipal wells, plus samples from the homes of cancer cases and three randomly chosen control homes. They analyzed the water for over one hundred substances, turning up some "solvent-like compounds" and nitrates. Once again, the levels did not indicate a health threat.

The scientists returned to the Voice of America radio broadcasting towers positioned outside of town. The initial county study had already considered the radio towers, but not everybody in the Department of Health Services felt satisfied with its conclusions. The level of doubt had been further raised by Dr. Louis Slesin, editor of *Microwave News*, a research newsletter published in New York City, who maintained that the Kern County health report paid "only scant attention to radiofrequency (RF) radiation." Slesin argued that "mounting data" showed that "low levels of non-ionizing radiation from power lines and radio waves play a role in the development of cancer." He cancer." He also stressed that waves of radiofrequency radiation could "act as cancer promoters. That is, they do not cause cancer, but encourage the development and proliferation of already-existing cancer cells—this would fit with the fact that no single kind of cancer is dominant among the children in McFarland."

The Department of Health Services research staff persuaded both the National Institute of Occupational Safety and Health and the EPA to monitor the broadcasting towers for harmful radiation. Neither agency turned up levels high enough to indicate health hazards.

Finally, the researchers directed their attention to what many observors had begun to consider the agricultural town's most likely source of contamination—pesticides.

McFarland stood only ten blocks wide, with no home located further than five blocks from the cotton, kiwi, and almond fields where spraying regularly occurred. Beyond these immediate contacts, the town was situated in the midst of the world's most productive farming region—and one of places most heavily treated with organic chemicals. Residents often complained of the chemical odors wafting across town during spraying season. For most people living in McFarland, pesticides were a fact of life—like the industrial haze of Pennsylvania steel towns or the coal dust shrouding Appalachian mining communities.

"The whole economy is based on agriculture," said Carlos Sanchez, who contracted cancer at the age of 20 in McFarland. "It's going to hurt the farmers if the study says it's pesticides. The health department is trying to make up its mind. Which is more important—profits or our health?"

In fact, the volume and variety of pesticide use in the McFarland region appeared staggering, if judged only by the official reports.

"I've seen the computer printout of pesticides used in this area over the past ten years," said Ed Dunne, who worked with the National Farmworker Ministry on the outskirts of Delano, "and it would stretch from this office to main street and back."

According to one account, more than 17 tons and 11,000 gallons of pesticides linked to cancer, genetic damage, birth defects, and reproductive problems were ladled upon the farming area surrounding McFarland from February 1979 to January 1983—a period encompassing the critical "window of opportunity" for harmful exposures.

Yet these figures could not reflect the full extent of pesticide use. California state law required official reports from farmers only for the pesticides known to be most dangerous, thereby bypassing chemicals whose health effects had not yet been calculated. Even the state agreed that the data base covering the Central Valley pesticides, like records for most agricultural regions, remained fundamentally flawed—thereby complicating, or as some argued, invalidating, any effort to pinpoint the role of pesticides in McFarland's health problems.

"Actual air monitoring for pesticides had not been done during the time period (January 1980 to January 1982) when a common exposure, if any, would have had to be present," noted the state's report. "Therefore, we must rely upon existing records of application. We recognize that this procedure might miss important pesticides not reported or might point falsely to pesticides used in a safe way."

Investigators could have tried to employ Pesticide Use Reports—the official records required by the state that noted the application of pesticides on farm lands. But these reports had one major flaw: The growers themselves filled them out. And Kern County employed less than one dozen inspectors to independently monitor the farmers' use of agricultural chemicals—in one of the largest, most productive farming regions in the world.

In addition to the Pesticide Use Reports' dubious value due to self-reporting, other factors may have made the records murky. The availability of black market pesticides (chemicals banned in the United States, but still available from Mexico in falsely marked containers) seemed to many critics to reduce the state's health study to a series of educated guesses—an even softer science than most epidemiologists liked to admit.

"There's a lot of ignorance in the world," said Chris Price, whose daughter had cancer. "There's so much we don't understand. I'm not sure that anybody's to blame. People out there on farms are using stuff banned years ago, and there's no way to do enough inspections. There are so many farms, so many farmers, so many chemicals being used, and so much greed, that you can buy and sell anything."

Fundamental questions about the migration and persistence of pesticide residues also remained unanswered. Not long after the state released its Phase II study, another report by the California Department of Food and Agriculture noted that cancer-causing pesticides had been entrapped in Stanislaus County's fog cover, thus contaminating test crops. Both the Department of Food and Agriculture and the Department of Health Services insisted that the pesticide residues trapped in the fog remained below levels harmful to people. Yet in truth, since the phenomenon had only just been discovered, nobody could authoritatively speak about the real health consequences of the pesticide-laden mist that often blanketed the farming areas—and traveled in unpredictable patterns across residential areas, such as McFarland.

Most important of all, no useful data existed to calculate the health implications of *combined* pesticide exposures over a short or prolonged period. Neither had researchers uncovered basic information regarding the human health effects of pesticides encountered in concert with nitrates in drinking water, electromagnetic fields, radio waves, or other toxic exposures.

"The pesticide DBCP showed up in our well," noted Rose Mary Esparza, "but they say it was below the level of danger. They know this from their rat tests. But the rats were exposed to DBCP alone. What if they gave the rats the DBCP, and then exposed them to all the other chemicals we are exposed to here in McFarland? What would happen to their rats then? Nobody knows. I think they'd end up dying."

Esparza's comments pointed directly to the researchers' worst fear: The source of McFarland's illnesses might actually be many sources—some incalculable combination of multiple toxic exposures, enhanced by diet, radio waves, nitrates in the drinking water, pesticides, or some other contributing factor, or factors, not even yet considered. In reality, the task of tracking down multiple culprits was

Rose Mary and Adrian Esparza. McFarland, California. Adrian had a muscle cancer removed from behind his eye.

comparable to identifying the names, shoe sizes, and Social Security numbers of several strangers with muddy feet who might have tramped past each other for a brief moment years before in the midst of a busy intersection. Given the limits of science, the chance of clarifying the long-departed convergence was probably zero.

Despite these formidable problems, the Department of Health Services pursued the pesticide issue. And it turned up one very provocative finding. The Phase II study demonstrated a tentative relationship between childhood cancer and parental exposure to pesticides. According to the study, "80 percent of the fathers of cancer cases, compared with 45 percent of the fathers of controls, stated that they had worked in the fields in the time interval between three months before pregnancy and the date of diagnosis of the child's cancer." Researchers made conjectures about fetal exposures, either through pesticides tracked into the household on the clothes or body of the field worker father, or by damage to his sperm. Yet this correlation was based on such small numbers (eight out of ten cancer cases, versus nine out of twenty healthy children) that the statistical linkage remained weak. And when state researchers reviewed the survey, they discovered that some people had been misclassified; for example, a man working on a farm as an administrator, who would have had minimal exposure to pesticides, had been misrepresented as a worker in the fields, where the exposure would have been great. After reclassifying workers in the original survey for greater accuracy, the researchers found the relationship between exposure and illness to be still evident, although considerably less striking.

Increased attention to local farming practices during the early 1980s also led researchers to identify four pesticides requiring further study: dimethoate, fenbutatin oxide, dinoseb, and dinitrophenol. Unsurprisingly, the state's continuing suspicions made many local farmers nervous.

"My interest, number one, is for the familes and the victims of this," insisted Warren Carter, president of Kern Farming Company, which grew almonds on seven hundred acres in the McFarland area. But Carter also expressed fears that state officials would "hang their hat" on pesticides—although in his opinion a single source of the cancers probably did not exist.

Don Riley, owner and manager of Ingram Kiwi Fruit, Inc., which operated in McFarland and nearby Delano, worried that public overreaction to pesticides "would be a disaster" for grape and almond growers.

Surprisingly, it was Jack Pandol, Sr., of Pandol Brothers, one of the largest grape growers in the country, and a long-time adversary of

the United Farm Workers union and their campaign against pesticides, who wholeheartedly endorsed the state health department's
continuing efforts. "I am glad that honest-to-goodness, good, consci-
entious people are making a study," affirmed Pandol. "I hope they
come up with an answer."

Despite the state focus on pesticides, the UFW remained profoundly skeptical. "They are not talking about the real toxicity levels
of these chemicals at all," complained UFW vice president Dolores
Huerta. "We just feel like the whole thing is a diversion tactic to keep
making people feel like something is being done."

Some McFarland families echoed the union's sentiments.

"At all these meetings," complained Tracey Ramirez, who had a
four-year-old daughter with cancer, "the growers say, 'Our number
one concern is for the children that are sick.' But not one of those
growers has ever come to me and asked how my child is."

In truth, disappointment over the study's failure to identify a
cause of the childhood cancers dismayed all sides. After more than
four thousand hours of work by the California Department of Health
Services, and considerable effort by Kern County investigators, the
mystery appeared no closer to being solved. Frustration on all sides
rose to new heights.

"You hope that you find a smoking gun in something like this,"
admitted Dr. Raymond Neutra, an epidemiologist with the Department of Health Services, "and we have not. Epidemiologists like us
have not had a very good batting average with these environmental
clusters."

Connie Rosales agreed with Neutra's assessment, though she
framed her disappointment in stronger language.

"I think the science of epidemiology needs a real kick in the rear
end," declared Rosales angrily. "We're getting into problems all over
the country and it's obvious there are environmental causations. The
science itself is just way behind."

Once the state released the Phase II study—completed at a cost
of over $650,000—the Department of Health Services braced for a
new storm of criticism. Dr. Tom Lazar, the author of the original Kern
County report who subsequently became its harshest critic, was still
working for the UFW—and once again, he drew widespread attention
by denouncing the state study. Most strenuously, Lazar opposed the
conclusion that arsenic was not the prime cause of McFarland's child
cancers. In support of the UFW campaign to drastically reduce pesticide use in the fields, Lazar maintained that arsenic could have been
left by sodium arsenate, a pesticide applied exclusively to grapes. Noting that empty containers of the pesticide had been discovered by the

railroad tracks outside of McFarland, he postulated that arsenic used on crops had somehow found its way into the water supply, though it showed up during tests only at trace levels.

Lazar's theory not only contradicted the state study; it also failed to explain why skin and lung cancers—the only kind of cancers associated with arsenical poisoning—did not appear among the nine cancers present in McFarland's children.

In August 1988, Lazar advanced yet another theory—one that several state researchers considered more plausible.

Lazar claimed that in the 1950s, agricultural businesses had dumped pesticides, mercury, arsenic, and other toxics in McFarland. Later, the area was covered over with soil from outside the town. Lazar identified this area as the tract home subdivision where half of the cancer kids had lived.

As new housing was constructed—with many occupants already settled in the neighborhood—the contaminated ground could have been dug up by work crews completing basements and foundations. This might have caused a short-term, high-intensity exposure, coincident with the timing estimated by the state health study. In fact, this theory mirrored events at Love Canal, where excavations of deep-set, contaminated ground temporarily worsened health effects.

The new theories, combined with mounting frustration over the state's supposedly negative health study, contributed even more to the growing misapprehension in McFarland.

Then shortly after Lazar's latest denunciation of the state report, the *Record* ran a story about Dr. James Fortson, a physician who worked two days a week in a Delano clinic, where he noticed numerous cancers of the tongue, neck, mouth, and sinus in adults from McFarland. Fortson's patients ranged in age from 40 to 73. But what struck the doctor's attention most was the strange rapidity with which their tumors grew.

"I saw a lady with a very small lump in the neck about the size of a pea," said Fortson. "She didn't want to have anything done on it over the Christmas holiday. When she came back two weeks later, she had masses on both sides of her neck. It went from the size of a pea to the size of two nectarines."

Fortson would not speculate why the normally slow-growing tumors would surge ahead so rapidly, but many McFarland residents drew an inference that new cancer-related problems had afflicted their community. Then in February 1988, doctors diagnosed another child in town with cancer. Residents shielded the identity of the latest child from the newspapers and television reporters, fearing that the family's private tragedy would be transformed into another public media event. Carlos Sanchez, who had cancer himself, would only divulge that the child was "a seven-year-old boy with a brain tumor." This additional case jarred the town. Another case had already been diagnosed in 1987—but not included in the study, because it fell beyond the investigation's time scope. And one more case would follow in 1989, one year after the study's completion. The official count stood at 13 cases with five deaths.

Equally disturbing came the news that McFarland was not the only town in Kern County that suffered from high, inexplicable rates of childhood cancer.

Early into the county's initial study, officials decided to track health records to determine if other small agricultural towns in the area evidenced problems similar to McFarland's. Surprisingly, they found the highest incidence of childhood cancer in Rosamond—a desert town, isolated from agriculture. Rosamond turned out to have nine children who had developed cancer between 1976 and 1986, with five fatalities—five times the expected rate for this town of 3,200. Five of the children suffered from brain cancer.

Then when state researchers began their investigation of Rosamond—long presumed to be environmentally pristine, given its apparent isolation from both agriculture and heavy industry—they uncovered a tremendous range of toxic contaminations. State staff found 35 toxic waste hotspots, including a 50-ton heap of toxic lead oxide, a 20-ton pile of lead waste from discarded car batteries, and 100,000 tons of petroleum coke containing potentially carcinogenic carbon compounds.

"I have never seen this much waste in one town before," said Kenneth Hughes, a hazardous waste specialist with the state's Toxic Substances Control Division. "The amounts and concentrations are alarming."

It seemed that wherever health officials focused their attention, more instances of contamination and disease suddenly materialized. The Department of Health Services began a year-long study in Rosamond, which ended up costing $500,000, but they could not determine the cause of the cancer cluster.

Back in McFarland, the quest for an answer was proceeding no better. In fact, some county and state investigators had begun to speculate that McFarland's cancer cluster mystery might never be solved.

"The way science works," explained Dr. Raymond Neutra, the state researcher, "you eliminate this, you eliminate that. And scientists feel very good when we are able to say, 'Well, we really have shown you it's not (this).' But then people say, 'But you haven't told me what it *is!* It's like a murder case: I've eliminated the butler. Don't you feel good? But, well, who did it?' "

Over time, the Department of Health Services, in conjunction with the state advisory panel, mapped out plans for the McFarland investigation's next phase. As Lazar had suggested, the houses in one particular subdivision could have been built on an old dump, where toxic compounds might have turned to gases and then entered the air after being stirred up during the housing construction. To evaluate this possibility, the state, along with the EPA, decided to test for sub-surface soil gas. Of course, the houses had been built years before, and any gases present would presumably have long disappeared. But the researchers still thought they might turn up something of interest.

The state agency also planned to conduct a new study to test the tap water in McFarland's homes to see if it could cause mutations of the DNA that could be verified in the laboratory. Genetic mutations of this kind might be responsible for the induction of cancers in children.

As the governor's "blue ribbon advisory panel" planned to meet for the fourth time, its members began to talk about a new idea. A study of Kern County alone had turned up an unexpected cancer cluster in Rosamond. What if they looked at the four major Central Valley agricultural counties surrounding McFarland—Kern, Tulare, Fresno, and Kings? Might they uncover additional cancer clusters?

"We could be having lots of places with high cancer rates," said Dr. Beverly Paigen, an epidemiologist who had conducted studies at Love Canal and now sat on the state advisory panel. "And McFarland happened to get noticed because it was just a little bit higher. I some-times worry that McFarland is not unique. If it's not, this is something with nationwide implications."

Yet in some ways, what the researchers feared most was that their intense, costly efforts might turn up even more cancer clusters—whose cause would ultimately prove as difficult to clarify as the mystery still plaguing McFarland.

"The longer we look and don't find a cause," admitted Dr. Lynn Goldman, the epidemiologist now heading up the state study in McFarland, "the more I'm convinced we're not going to find one."

But in McFarland, the studies continued.

The Struggle for Certainty
Science Runs Up Against Its Limits in Toxic Towns

Glen Avon, California. Residents leave empty bottles of purified water in front of their homes. Government trucks replace them with a fresh supply.

Nobody could be certain about what happened in Glen Avon.

In the early 1980s, residents of this small Southern California town began to confide in each other about a puzzling array of illnesses suddenly afflicting their familes. At the supermarket, over backyard fences, in the waiting rooms of doctors' offices, the mothers of Glen Avon exchanged stories about their children's lingering cases of asthma, dizziness, headaches, blurred vision, and persistent colds. Some even remarked on the sudden rash of cancers striking down their pets.

"For two years," recalled Penny Newman, a longtime Glen Avon resident, "my son went through medical tests for neurological problems. He had headaches and seizures. Nobody knew why." Similar stories from other mothers disturbed Newman even further. She haunted the local doctors' offices and hospitals, searching for an explanation for her son's perplexing ailments. "It wasn't until the doctors couldn't find anything else," said Newman, "that I mentioned, 'Well, we do live near this hazardous waste dump. . . .' "

The hazardous waste site that Newman offhandedly brought up was the Stringfellow Acid Pits, a huge dump for liquid toxic wastes erected in the Jurupa Mountains overlooking Glen Avon. The Stringfellow Acid Pits contained millions of gallons of solvents; heavy metals; pesticides; sulfuric, nitric, and hydrochloric acids; and a vast assortment of other chemical liquids discarded by industry and agriculture. But Newman understood that proximity to these hazardous wastes was not the same as exposure. Only gradually did she begin to wonder if the toxic wastes seemingly contained by the Stringfellow Acid Pits could have somehow reached the people in her town.

In fact, there were numerous possibilities.

When heavy rains hit the area in 1977 and 1978, the liquid toxic waste dump abruptly overflowed. Several million gallons of chemicals washed down from the foothills above Glen Avon to mingle with the rainwater already gushing through the suburban streets. Although the toxics streamed across the city's sidewalks, gardens, school yards, and playgrounds, almost nobody initially perceived them to be a health threat. These wastes had been diluted in the massive flood; in fact, their presence was scarcely noted by most people. "We were worried about mud in our homes," recalled Newman, "not toxic chemicals."

But over a protracted period of time, other opportunities for exposure to the dump's chemicals also occurred.

Before and after the floods, children playing in the hills around the dump could have come into direct contact with its contents. Windborne gases and chemical particles might have been swept down from the dump and into the city streets. Soil and water contamination also persisted long after the dump's overflow. Any of these chemical contacts would have involved low levels of exposure. The chemicals' routes and duration remained extremely uncertain.

Yet over time, Penny Newman began to feel more convinced that her town's chronic, diverse, low-level exposures to the toxic dump had indeed compromised the health of her son and numerous other people. In fact, in her mind, the link between the town's massive toxic exposure and its numerous illnesses simply stood to reason.

Proving her case, however, was entirely another matter.

—   —   —

In the beginning, Penny Newman's best hopes to prove that her town had been made sick by toxic wastes hinged on the science of epidemiology.

Most narrowly defined as the study of epidemics, epidemiology really covers the broader ground of searching for associations between factors that can cause illness and the illness' occurrence. Epidemiology's first hero, Dr. John Snow, earned his place in medical history by puzzling out the origins of the London cholera epidemic of 1854—saving thousands of lives through his shrewd deductions.

Snow observed that one particular London neighborhood ranked significantly higher in cholera cases than the rest of the city. While strolling through the neighborhood, the doctor surmised that many of its residents drew their water from the Broad Street pump— whose source he traced to a part of the Thames River where other city dwellers dumped raw sewage, including excrement from cholera victims. Snow immediately grasped that the contaminated water recycled the disease throughout the city. He took action to stem the epidemic.

In handling infectious diseases like cholera, or by constructing cause-and-effect associations between habits like cigarette smoking and cancer, or high-fat diets and heart disease, Dr. Snow's twentieth-century epidemiological descendents have proven eminently successful. Modern epidemiologists have halted the spread of typhus and typhoid, tracked the course and transmission of AIDS, and identified the cause of Legionnaires' disease. But in the field of environmental epidemiology, the study of how factors in the environment affect human health, these same investigators have stumbled blindly through an array of scientific, statistical, and procedural mazes— many of their own making. Snow's commonsense solution to the cholera epidemic plaguing nineteenth-century London, in which a single contamination caused a single disease, now seems an elementary feat compared to the rigors of solving today's toxic mysteries, in which countless chemicals may stream into society through various and complicated means.

In Glen Avon, the situation proved exactly how daunting environmental epidemiology could be—for *everybody* involved.

Glen Avon's saga actually began in 1956, when businessman James B. Stringfellow, Jr., opened the Stringfellow Acid Pits in the foothills above town. Assured by geologists that the Jurupa Mountains' highly impermeable pyrite rock would prevent wastes from percolating down to the groundwater, Stringfellow dug 20 enormous trenches and filled them during 18 years of operation with 30 million tons of highly toxic heavy metals, carcinogens, mutagens, nerve toxins, and other wastes generated by industry and the military.

Then came the rains.

In 1969, torrential downpours at the edge of this desert community caused the acid pits to overflow and rush down into nearby Pyrite Creek, which in turn gushed into the streets of Glen Avon. The people in town had no way of knowing that the water flooding their basements, yards, and wells also contained hazardous chemicals from the dump above their homes. They simply cleaned up the water, dried out their possessions, and went on with their lives.

Three years later, health officials discovered that the well supplying drinking water to the 670 children attending the local elementary school had been contaminated with dangerous chemicals. Nobody could determine when the contamination had occurred. The health department closed the well.

"My two girls were in the elementary school when they found the contaminated well," recalled Shirley Ann Neal, another Glen Avon resident. "We got a note saying they should bring bottled water to school with them to drink. I figured the well's got bacteria or something. So we just sent jars of water to school with them."

But bottled water at the school did not protect Glen Avon's children from other chemical exposures. Many kids routinely explored the hills and caves above town, even venturing into the dump. The more adventurous kids pirated the company's maintenance boat and rowed across the ponds of toxic wastes. They skimmed foam from the surface, covering their faces with bubbly white and yellow beards.

After investigators located the contaminated well on school property, the county health department ordered James B. Stringfellow, Jr., to make certain his waste ponds would never again overflow into Pyrite Creek and jeopardize the community's water supply. Stringfellow consulted with his engineers and lawyers. The next day, he closed up shop. Refusing to accept further shipments of chemical wastes, Stringfellow erected a containment dam around the waste ponds to keep his standing load in place.

In this sense, the Stringfellow Acid Pits ceased operation in 1972. From another perspective, the real problems had just begun.

Heavy rains throughout 1978 threatened to rupture the dump's new containment wall. To prevent a massive disaster, the Regional Water Quality Control Board, along with the County Board, decided to engineer a "controlled release" from the ponds. They punctured the dam at a strategic point, letting loose eight hundred thousand gallons of water from the hazardous waste ponds into Pyrite Creek.

Once again, the creek overflowed, flooding some one hundred gardens, yards, and homes in Glen Avon, and pouring into the school yard. County officials did not notify residents or school administrators about the controlled release. But shortly after this time, some parents noticed that the soles on their children's shoes began to disintegrate when they played around the creek.

The next year, the California State Department of Health Services discovered DDT in a three-acre district of the town. Workers decked out in protective clothing excavated the contaminated soil; but Glen Avon residents complained that workers temporarily piled it in the center of town in a mound running two hundred feet long and ten feet high. The DDT-laden soil remained there for two weeks, being blown whichever way the wind decided.

The spectacle of workers in protective clothing digging up the contaminated soil drastically raised the town's anxiety level. A new community group rapidly formed, with its members demanding the answer to one question: Had any of the town's several low-level toxic exposures compromised their families' health?

Most active among the aroused citizenry was Penny Newman. For years, Newman had been the community spark plug—active in the PTA, Little League, and Cub Scouts, and employed as an instructional aide at the elementary school. The local West Riverside Businessmen's Association had even named her "Citizen of the Year." As Newman grew increasingly concerned about the toxic chemicals that had filtered through her town, she approached the problem with the same levelheaded zeal that had served her well in the past.

"It was just another thing to work on," she explained to us when we visited Glen Avon. "Instead of trying to get a stop sign at an intersection to protect the kids, I was going to work on the hazardous dump. I went into this thinking we'll have a community meeting, show the county officials our concerns, and they'll correct the problem. We really believed that. And now, more than a decade later, we're still here trying to get it done."

Word had already spread through town about debilitating headaches, strange skin rashes, muscle weakness, chronic nosebleeds. Newman's own son suffered from muscle pains and seizures. Her neighbor, Shirley Ann Neal, complained about the strange illnesses affecting both her husband and children. Even the family's horses, who had stood in three feet of water during the flood, had all developed skin problems—and two died from cancer. Another neighbor, Martha Rodriquez, told Newman about her 3½-year-old son, Harvey.

Doctors had found the boy's liver and spleen to be enlarged; internal bleeding had required an emergency blood transfusion. Harvey had leukemia.

"We didn't know anything about toxic wastes until Harvey got sick," admitted his mother. "But then we started seeing things in the paper about chemicals in the water that could cause cancer. It was unreal. I've lived here for 13 years, and I didn't know those pits were directly above our town."

When the county health department finally acceded to the new community group's demands, and agreed to conduct blood tests on residents, many people felt enormously relieved. "We were extremely excited," said Newman. "We'd be getting some answers."

The health officials tested a number of Glen Avon residents who had taken ill for DDT in their blood. Some registered positive, some did not. Then the investigators tested a control group composed of presumably healthy people living outside of Glen Avon: same result.

In truth, the county's tests proved nothing. DDT comprised only a small part of the chemicals contained by the dump. Besides, a huge number of Americans throughout the country had DDT in their bodies. Glen Avon residents who tested positive for the pesticide could not know if it had caused their headaches or other health problems, while the people who showed no traces of DDT in their bodies still knew nothing about the health effects of hundreds of other chemicals that had washed through town. The county's efforts turned out to be as amateur a science experiment as could possibly be designed.

"The county wasn't interested in testing properly," complained Newman, whose celebrated civic virtues were beginning to flare into intense cynicism about government. "And that made sense. The people in charge of the blood tests were the same ones who gave String-fellow the permit to put the dump up there in the first place. They had made the decision to have the 'controlled release' of water from the chemical ponds that ran through the town in 1978. They weren't interested in finding something bad going on here, because if they did, they were to blame."

Newman and her neighbors wrote off the poorly designed study as the product of local incompetence and an irredeemable conflict of interests. Yet most people in Glen Avon still believed that a better-crafted investigation, overseen by genuine experts in the field, would produce drastically different results. "We were still in the mentality," recalled Newman, "that if we showed them the problem and proved the facts, the government agencies would respond and help us out. If we could only get a good health study done here, something would get done."

In 1983, the opportunity arose unexpectedly to undertake the largest and most exhaustive health study ever conducted at that time in a chemically contaminated town. A federal court had recently found the Reagan administration's EPA hazardous waste director, Rita Lavelle, guilty of perjury for her testimony regarding the government's lax attempts to clean up the Stringfellow Acid Pits. Lavelle's boss, EPA chief Anne Gorsuch Burford, was also cited for contempt of Congress in her own dealings over Stringfellow. Suddenly national attention focused on Glen Avon's problems, and government funding for the study instantly materialized.

Much to the delight of Glen Avon residents, the State of California, and not county officials, was charged with overseeing the project this time. The State Department of Health Services contracted with UCLA's School of Public Health to design a study, with its research team headed by two respected scientists, Dr. Dean Baker and Dr. Sander Greenland. The state also assembled an advisory board to assist with the project's initial design. Its members included both Penny Newman and Dr. Beverly Paigen, an epidemiologist whose pioneering work at Love Canal had led her to collaborate with community groups throughout the nation seeking medical evidence of their town's contamination. Paigen's sterling reputation as a "people's epidemiologist" qualified her as one of the few scientific experts trusted by the residents of many toxic towns.

In 1983, twelve trained interviewers spread through Glen Avon, reaching 606 households to obtain information on 2,039 residents. The all-female staff spent 1½ hours at each house, usually talking to the "female head of the household" deemed most likely to be acquainted with the health histories of the people living under her roof. The interviewers' questionnaire ran 125 pages, covering everything from eating habits to urinary patterns. A study team also reviewed death certificates from the area and official medical records.

The thrust of the study was to compare the health of people exposed to Stringfellow's toxics with other people who had not been exposed. To this end, the researchers divided their interview subjects into three groups.

The people living close to the dump, or to the creek that had flooded the town, were designated as having the greatest likelihood of exposure to toxic chemicals. Residents of West Glen Avon, across town from the dump and the creek, were considered to have a "small" exposure risk. And people living in Rubidoux, a small town three miles south of Glen Avon, separated by a row of hills, were deemed "non-exposed."

Three years later, in 1986, the interviewers had finished their jobs, the data had been entered in the computers, the statistical analysis completed—and Dr. Dean Baker and Dr. Sander Greenland were able to pore over it all and report their findings to the California State Department of Health Services in a five-hundred-page tome.

Over those three years, Penny Newman had unsuccessfully battled local, state, and federal government agencies in an attempt to win medical care for Glen Avon residents and secure a commitment for the complete cleanup of the Stringfellow toxic waste dump. Now she hoped that the results of the health study would provide the necessary ammunition to victoriously conclude her fight.

But Newman did not get an opportunity to review the study until minutes before the Department of Health Services held a press conference to announce its findings. And what she read in those few minutes stunned her.

The state health officials reported that the three study groups showed no difference in cancer rates, miscarriages, birth defects, or deaths. And although 23 different symptoms of illness were reported at uniformly higher rates in the population designated to have the greatest likelihood of exposure to toxic chemicals, the officials discounted these symptoms for a variety of reasons that immediately angered Newman and would soon outrage other Glen Avon residents.

"I was so embarrassed at the press conference," recalled Newman. "I felt like I had let my community down. Here I had participated in this study, and I wasn't even given an opportunity to look through it so I could respond. We were caught by surprise, without time to pull anything together."

Somehow Newman weathered the press conference. The follow-up community meeting proved tougher. That evening three hundred Glen Avon residents piled into the local elementary school to hear the state scientists explain the study's results.

"Nobody could make heads nor tails out of what they were saying," sighed Newman. "They hung up these graphs at the front of the room and talked about things like 'age overlay' and 'statistical methodology' to the point where nobody had any idea what was going on. It was total mush."

The next day, the newspaper headlines announced with strident clarity that no adverse health effects from toxic wastes had been found in the Stringfellow health study.

Two weeks later, Penny Newman and scientist Beverly Paigen joined forces to conduct their own community meeting.

"We knew that we had to respond to the study very quickly," said Newman. "People had such a feeling of disillusionment; it was so devastating."

Once again, Glen Avon residents returned to the elementary school for another discussion of the health study. Most of the media skipped this presentation, missing a very good show.

Newman and Paigen began the evening by posting a series of graphs at the front of the room that closely resembled the ones used by the health department two weeks earlier when it attempted to illustrate the community's rates of expected-versus-actual diseases.

"Our graphs showed the amount of bullshit we had expected from the state," explained Newman, "and the actual level of bullshit we'd received. The second graph was a whole lot higher than the first."

After the laughter died down, Paigen moved on to her own interpretation of what the study said—only in language that everybody could understand. "Then," recalled Newman, "Bev Paigen explained what the study really *meant.*"

In Penny Newman's view, the only thing the study truly indicated was that the people of Glen Avon had been deceived and abandoned by the State of California.

Many people in town *knew* they were sick, and despite the bland denials of the health department, they blamed the waste dump. The people of Glen Avon had participated in good faith with the health study in hopes of obtaining medical care for their illness, monitoring their families for future problems, and finally getting the dump site cleaned up.

But now after three years, the study provided no answers or help whatsoever.

As Penny Newman and Dr. Beverly Paigen stood at the front of the Glen Avon elementary school, gazing out at the faces of hundreds of angry, disillusioned citizens, they could only ask themselves one question: What went wrong?

- - -

Environmental epidemiologists everywhere operate under one fundamental, unremitting restraint: They cannot conduct experiments on human beings. Instead, researchers of environmental effects must be content with observing what has already happened to people—and then thrash through murky intricacies and muddy contradictions to arrive at their clear-cut conclusions.

Naturally, this limitation makes sense. Nobody would argue that, for the sake of speed and clarity, scientists should expose half the residents of some town to a known toxic chemical, leave the other half unexposed, and then compare the effects.

Yet at first glance, Glen Avon seemed to afford the state's epidemiologists precisely this kind of opportunity—thanks to the toxic wastes pouring down from the Stringfellow Acid Pits. Like laboratory mice destined to serve as the unwilling subjects in some high-risk experiment, Glen Avon residents found themselves neatly divided into eminently studyable groups of the "exposed" and the "unexposed."

The truth, however, proved a great deal more complicated.

In the environment of the laboratory, scientists find no difficulty in distinguishing exposed groups from control groups. Rodents, rabbits, or other animal captives can easily be isolated from the vagaries of everyday existence under the constant supervision of the researchers. In the laboratory, every animal eats the same food, maintains the same exercise regimen, works the same job, endures the same amount of daily stress, earns the same income, abstains from smoking and drinking, doesn't take drugs on the sly, and goes to the same doctor for health care and diagnosis of disease. Most important, the scientists know exactly which animals have been exposed to the chemical they intend to test, and which have been spared the experience. Every animal in an "exposed" group is unequivocally exposed; every animal in the "unexposed" group is definitely not exposed.

Glen Avon's elementary school. The channel for Pyrite Creek runs alongside the school playground.

In Glen Avon, epidemiologists operated with far less certainty.

To conduct their study, the scientists needed to come up with an "exposed" and "unexposed" group for comparison. But their choice of the neighborhood closest to the dump as the exposed group presumed that *everybody* living there had come into contact with toxic chemicals. Clearly, some had. Just as clearly, some were "less exposed," and some were not exposed at all. The people with little or no exposure might have more recently moved to the area, or they could have been away from town during the floods, or they simply might have had less contact with the chemicals than their neighbors because of their daily habits. Yet everyone with a home address in the "exposed" area was deemed by the study as having had an identical exposure.

The same inconstancies plagued the rest of the study's groups. A neighborhood located further away from the waste dump and the creek was selected as the "less exposed" group, and another neighborhood, even more distant, was deemed "unexposed." Most individuals in these groups did indeed reflect their collective characterizations. But some unknown number of people who didn't actually live near the wastes had probably been exposed to them through various uncharted routes.

Some children from the "less exposed" area, for example, might have played at the waste pits, sloshed through the flooded streets, or rambled around the chemically saturated school yard. Their parents might have worked downwind from the fumes blown off the ponds; or during their lunch hour, they could have breathed in the dust from the mounds of DDT-laden soil piled in the middle of town. Teachers might have lived outside the "exposed" neighborhood, but they nevertheless spent five days every week at the school, next to the flooded creek; their contact with the chemicals could have been greater than somebody who resided in the high exposure neighborhood, but worked out of town. Some residents of West Glen Avon, the "small exposure" neighborhood, probably visited friends and relatives near Pyrite Creek, and even helped them mop up their flooded homes— thus, increasing their own chemical contacts. If families from Rubidoux, the "unexposed" community across the hills from Glen Avon, visited their relatives during the floods, or their teenagers trekked up to the dump with their Glen Avon cousins for an afternoon of illicit adventure, they, too, deserved to stand among the "exposed" population. But the study's protocols declared them all unexposed.

In truth, it was as if some mischievous gnome had slipped into the laboratory at night and exchanged an indeterminate number of mice from the unexposed control group for the mice residing in the cages marked "exposed." Of course, if scientists conducting a laboratory experiment found out about the gnome's antics, they would have thrown up their hands, considered their basic data hopelessly muddled, and started all over. In Glen Avon, the researchers faced the same situation—but they came up with firm scientific conclusions.

Even from the start, the study's designers recognized this problem. According to Dr. Dean Baker and Dr. Sander Greenland, the UCLA professors engaged to oversee the research, their metholodogy had "assigned equivalent exposure status to every individual within the same study area, even though the individuals certainly did not have the same level of exposure to toxic waste. The probable effect of exposure miscalculation would be to reduce an apparent association between 'exposure' and health outcome rates. . . . It is possible that many individuals in the Glen Avon study actually had negligible exposure to toxic wastes. . . . This would tend to obscure true differences in health outcome rates due to exposure."

Other scientists supported this conclusion, noting that these kinds of problems affected not just the Glen Avon study, but practically all investigations straining to establish a link between community health effects and toxic chemicals. "Every time we use an imprecise measure of exposure, like what block you lived on or what job you had," explained Dr. Lynn Goldman, head of the environmental epidemiology section of the California State Department of Health Services, "it decreases the precision of the study. It makes the tool we probe with very dull."

Yet, given the practical limitations of epidemiology, combined with Glen Avon's complicated history of toxic exposures, the researchers had arguably done the best job possible, forging an imperfect, but necessary compromise over the study group classifications.

Members of the community, however, wondered why the researchers simply didn't pay more attention to *their* advice.

"We had been asked to participate in the development of the study protocol," lamented Penny Newman, "because we lived in this community and knew better than anyone else what exposures might have happened when and where. The researchers wanted to know what our concerns were, what populations should be looked at, why would such-and-such an area not be a good control. We knew that new houses had been built near Pyrite Creek and that people who lived in those homes probably had lower exposures, but were considered as part of the high exposure group. And we knew about families who lived far from the dump and the creek, whose kids went to school near the creek—and were driven there during the floods by parents who waded through the water with them to the front door. We knew that many people who lived in West Glen Avon, the 'small exposure' group, went to work in the high exposure area every day."

As a member of the advisory board, Newman had offered dozens of recommendations to the researchers for shaping their study. But she never actually saw the protocols governing their work until three years later when the study was completed—and then, only hours before the press conference announcing its profoundly negative conclusions.

"When they finally let us see how they actually picked the people and areas to study," chided Newman, "it reflected everything we had pointed out in the advisory committee that they shouldn't do."

Coming up with the right groups to study was not the only serious problem facing the Glen Avon researchers. Epidemiology is fundamentally a numbers game, in which the laws of statistical probability reign supreme. But in order to play the game fairly, large numbers are required. When only small numbers prove available, the calculations inevitably misfire.

This concept can be explained best with a simple toss of the coin.

When you flip a coin in the air, it will fall to the ground with a 50-50 chance of landing on heads. If the coin lands on tails three times in a row, you could not infer that some mysterious force was affecting the outcome. You would simply need to flip your coin a few more times; then the sheer coincidence, random chance, luck—what you will—of the tail-landing toss would begin to even out with an approximately equal number of heads-landings. If you could muster the patience to toss the coin one thousand times, you are assured a near-even split between heads and tails. If not, suspect the coin.

In studying the connection between disease and its causes, scientists must also consider the possibility of random chance.

If a town of seven thousand people turned up two cases of cancer, and another town the same size had four cases, the statistics could not tell you if there was any real difference between the two communities' cancer rates. Coincidence alone might account for the doubling of cancers. In this case, the small population sample is the pathological equivalent of the coin toss that lands three times successively on tails. But if you compared a city of six million people afflicted with two hundred cases of cancer against a comparable city of the same size with four hundred cancer cases, you could be reasonably certain that the doubling wasn't due to chance. With numbers this size, an epidemiologist could analyze the situation and arrive at a 95 percent certainty that the cancer difference indicated some irregularity. And then the real work of tracking down the cause would begin.

According to the rule books of science, epidemiologists can only point to a "significant" difference in the level of disease in a community when their statistical analysis says there is more than a 95 percent chance that coincidence could not account for the difference. Anything less—say, an 85 percent chance that coincidence was not the factor—means that the scrupulous researcher must cooly state that the statistics don't confirm any significant increase in illness. And it takes very large numbers to achieve this level of 95 percent statistical certainty.

In Glen Avon, the 2,039 research subjects were split into three groups for the purpose of statistical analysis. Because these groups were so small, epidemiologists required a vastly increased cancer rate among the exposed group before their statistical analysis showed a 95 percent chance that the observed increase wasn't due to mere coincidence.

"If 30 to 40 percent of the cancers in Glen Avon were caused by chemical exposure," explained Dr. Raymond Neutra, the Department of Health Services epidemiologist who headed up the state's investigation, "the study would not have discovered it. You'd have to have a tripling or more in the rate of cancer before the study would detect it. The power of the study was not great because the number of people and illnesses we were working with were so small. There would have to have been a pretty major health effect for us to have found it. That's the limitation of the science."

The lag time between exposure to cancer-causing chemicals and the onset of disease also confounded the ability of the study to pick up any link between toxics and cancer. Cancer may take as long as 10 to 30 years to develop in the wake of a toxic contamination. The major exposure in Glen Avon probably occurred in the 1970s, much of it in the late 1970s. Therefore, most chemically induced cancers would not have occurred by the time of the 1983 study.

"Certainly," said Dr. Neutra, "if we repeated the Glen Avon study in 15 years, we might see a significant effect then. The cancers are more likely to show up in the future."

Statistical complications played complete havoc when it came to investigating the health problems of Glen Avon's children. After dividing the population into three groups, researchers found that there simply weren't enough kids to conduct an accurate analysis. So they limited their findings to residents over 21 years of age. While the parents in town focused their worries on the health of their children—and while children's developing bodies make them the most likely candidates for toxics-induced illnesses—the Glen Avon health study said nothing at all about their health. This may have made mathematical sense, but it enraged many Glen Avon parents, who viewed the omission as a major defect in the study.

Martha Rodriquez, whose son, Harvey, had leukemia, responded to this beguiling numbers game by falling back upon her own fears and suspicions.

"They keeping asking me, 'What proof do I have that Harvey's leukemia is from the chemicals?' " said Rodriquez. "I can't be *certain*. But the water had all these chemicals in it that can cause cancer. I wouldn't have another child here."

Without doubt, the health study's accuracy had been compromised by a profound combination of uncertainties. But instead of publicly admitting the study's inherent flaws, the scientists appeared to be indicting the community for its own overwrought sensitivities and hyperbolic fears.

First of all, there was the matter of "recall bias."

Recall bias refers to the indisputable fact that people who are worried about their health tend to remember, and sometimes exaggerate, their ailments more than people who have no health concerns. The Glen Avon study did come up with statistically significant differences in many illnesses. Ear infections, bronchitis, asthma, the chest pain of angina pectoris, ulcers, and skin rashes all were elevated in

Harvey Rodriquez in front of the EPA Superfund cleanup site at Glen Avon.

the exposed part of town. And eight other symptoms, including blurred vision, ear pain, daily cough, nausea, diarrhea, dizziness, and frequent urination were also at a higher level. Because of this broad-based elevation of illnesses and symptoms, the researchers immediately suspected recall bias. They concluded that people living near the dump tended to remember more symptoms and relate them to the interviewers precisely because their concern about the dump had primed their memory.

Actually, the researchers had anticipated the problem of recall bias—and constructed a method that they hoped would neutralize it. The Glen Avon researchers had drawn upon the example of Boston epidemiologist Dr. David Ozonoff, who had studied the health effects of hazardous waste exposures in Lowell, Massachusetts. As a hedge against the community's enhanced concern about illness, Ozonoff asked his research subjects if they believed the air or water in their neighborhood made them sick. Then he eliminated from his study every person who answered yes. Nevertheless, among the people who declared that they believed environmental pollution had not harmed them, Ozonoff found increased rates of earaches, sore throats, and frequent colds—and the closer the respondents lived to the dump, the more their symptoms increased. According to Ozonoff, the elevated rates of illness in people who believed they were not made ill by toxic chemicals eliminated any effect of recall bias from his study.

Glen Avon's researchers attempted to sort out recall bias in a similar way. They searched for a group, like the Lowell unbelievers, who did not worry about their proximity to hazardous wastes. Unfortunately, they couldn't find one. In Glen Avon, West Glen Avon, and even "unexposed" Rubidoux, virtually every household believed that toxic wastes had jeopardized their health. As a result, Baker and Greenland modified Ozonoff's approach, dividing people in their own study into groups of "highest perception of threat" and "other than highest perception of threat"—the very worried versus the merely worried. And they found that the very worried identified more symptoms than the merely worried. Because of this elevated rate of symptoms in the very worried, the researchers concluded that enhanced recall had affected the results of their study. The elevated rates of illnesses in town were discounted as a memory problem, not a problem of diseases.

Once again, Penny Newman thought this conclusion made no sense.

"It's not like this community is a hotbed of hysterics," she said, citing another part of the study that showed no increased level of anxiety or depression in the exposed population. If anything, Newman claimed, people in her town tended to minimize their symptoms. "It's like the smog in Los Angeles," she said, "it's a way of life to have burning eyes—you take it for granted and don't even mention or think about it anymore. In the same way, people in Glen Avon have become used to being near this dump. We don't think so much about it, but like the people in Los Angeles, we still suffer from the symptoms of pollution."

Glen Avon's researchers also objected that the community's long list of health problems sounded doubtful since the majority of symptoms—including eye irritation, blurred vision, ear pain, dizziness, nausea, and diarrhea—were "self-diagnosed." And when the investigators reviewed the residents' medical records, the physicians' reports failed to back up their complaints.

When Penny Newman heard that the researchers discounted symptoms that were not backed up by official medical records, she minced no words. "That's just hogwash," she said. "I know the people in this community. Bloody noses aren't reported in medical records because most people here don't go to the doctor when they have a bloody nose. And those that went to find out why they were having *so many* bloody noses got tired of hearing, 'I don't know what's wrong.' We've gotten tired of going to doctors with so many problems and finding no explanation. People around here can't afford doctor's visits where nothing happens."

Glen Avon mother Shirley Ann Neal also regarded these objections skeptically.

"All kids tend to have infections and bloody noses," admitted Neal, "I understand that. But all three of my grandchildren have bloody noses and colds and ear infections and coughs—all of the time." Neal didn't feel she needed to consult the health study to prove that something had gone wrong in her community. "There's two babies on the way in this family," she said. "I'm going to get out of here before we have those babies exposed."

Finally, the researchers doubted the Glen Avon residents' symptoms and diseases because there were so many *different* complaints. "For example," their study stated, "it is reasonable to believe that exposures to toxic chemicals in surface water or soil may have caused skin rashes." But no toxicological mechanism existed to explain the additional presence of nausea, ear pain, blurred vision, and a plethora of other symptoms. Once again, the inexplicable symptoms led the researchers to think that increased recall of disease was more of a factor in Glen Avon than toxic contamination.

"It's like they're calling us hypochondriacs," complained Penny Newman. "This community was exposed to so many different chemicals in so many different combinations and by so many different routes. And then every individual has their own health strengths and weaknesses. From my perspective, it's logical that there would be a lot of different illnesses at higher levels. It's not as clear-cut as they want—like when the scientists investigate exposure at some factory to a single chemical. Our situation is much more complex.''

Newman was correct in stating that the epidemiologists preferred cleaner, less compromised studies. Like Dr. John Snow of London, the nineteenth-century pioneer who investigated cholera, modern environmental researchers perform at their peak when attempting to establish the toxic effects of a single substance causing a single disease.

Industrial workers exposed to asbestos, for example, demonstrate high rates of mesothelioma, a lung cancer rarely found in people without asbestos exposure. Although it required years of painstaking research, epidemiologists eventually proved the connection between the toxic substance and the disease. The same held true for a rare liver tumor called an angiosarcoma, which was found at elevated rates among factory workers producing the chemical vinyl chloride—and later among people living near the factories. The tight association between the seldom-seen angiosarcoma and the high exposure to vinyl chloride made it possible to dismiss the vexing questions of recall bias, self-reporting, and multiple symptoms.

But in towns like Glen Avon, where uncertain combinations of vastly different chemicals had swept through the community in various strengths and doses—and may be capable of causing a wide variety of related or unrelated illnesses—the epidemiological answers remained singularly remote. Instead of reliable data or relief, the final reports of the scientists, who had been brought to town at an enormous cost to the taxpayers, produced only huge amounts of frustration, grief, and rage.

- - -

''We made a big mistake,'' admitted Penny Newman. ''We thought we needed tests, scientific proof. If only we had the data, we were sure the officials would take care of our health and clean this place up. What we didn't understand when we embarked on the health study was that the state already knew what the results would be. They knew that their science would end up showing no significant health effect.''

To Newman, the facts in the case of Glen Avon were indisputably simple.

"People were exposed," she summarized, "and it was not good for their health."

But despite all the questions about the statistics, the exposed and nonexposed populations, the lack of available children to study, and the doubtful significance of recall bias, the State Department of Health Services had reported to the people of Glen Avon, with perplexing certainty, that the toxic exposures in their town had caused them no significant health problems.

It wasn't just the health study's galling conclusions that made many Glen Avon residents rail against the state government.

"This whole science thing is political," insisted Newman. "Even if it was scientifically possible to conduct a study here that would show the effect the dump had on our health, they would design it to not show anything. They manipulate the facts. And they design studies to create the right headlines for them politically."

By the time the health study was finished, the people working through Newman's organization, Concerned Neighbors in Action, started to regard almost every move by the state with the utmost suspicion. Many people wondered, for example, about the state-sponsored trucks that pulled into town each week, loaded down with five-gallon jugs of purified water. The water was distributed without charge to residents for drinking, cooking, and washing.

"They tell us the water in Glen Avon is safe," said Shirley Ann Neal, "and yet they provide us with free bottled water. Two and two is not adding up here."

One of the most persistent and convoluted controversies between the citizens' group and the state began with a rather straightforward error by the health department.

On March 10, 1986, Dr. Raymond Neutra chaired a meeting of his department's Stringfellow Planning Committee to review the findings of the Glen Avon health study conducted by scientists from UCLA. The meeting's minutes, intended for internal use only, contained a sentence reading, "Excesses were found for the incidence of all cancer and for skin cancer." Clearly, the statement reflected a transcription error, since the study had found no such thing.

But Penny Newman managed to get a copy of the department's minutes, and when she read the statement about the increased cancer rates, she decided that there had been a cover-up within the Department of Health Services, led by Dr. Neutra. Of course, Neutra explained the error, but Newman didn't believe him. And when Glen Avon's Concerned Neighbors in Action published their "Stringfellow Acid Pits Fact Sheet" for other residents and the media, they asserted publicly that the health study had shown an increased incidence of cancer.

By 1992, after years of battling with the state health department, Newman had only slightly softened her position on the mistaken minutes.

"Dr. Raymond Neutra has the burden of explaining the mistake," she told us. "What it did prove was that the state is not infallible, that there were excesses that couldn't be explained—and most of all, that their doubletalk misleads people."

Neutra seemed an unlikely villain. In fact, he readily admitted the mistakes made by his department in dealing with the community. And he conceded privately that the health studies might have failed to pick up a doubling or even quadrupling of various illnesses. More generally, Neutra decried the state of the nation's contamination that led to problems, like those affecting Glen Avon, in countless other communities.

"It's a terrible thing that waste practices have put communities in this place where they don't know whether or not they're suffering from unusual diseases due to toxic exposures," said Neutra. "And nobody can give them an answer. Yet there's a commonsense element to all of this, too. *Of course* you don't want pesticides in your drinking water, and *of course* you don't want chemicals running down the drainage ditch in the middle of your community. We shouldn't wait until we can count the bodies in the streets before we recommend action."

And yet for Penny Newman it was this very reasonableness personified by Dr. Neutra—and shared by virtually all his scientific colleagues—that caused her to feel most disheartened and suspicious. It wasn't just a matter of personalities, she seemed to be saying. The system itself, regardless of the best intentions of everybody involved, simply wasn't working. Industry and agriculture could contaminate, waste dumps could leak, communities might suffer from increased disease, studies might be commissioned and completed—and nothing would really change.

This sense of systemic collapse was further intensified in Newman's community by the state's slippery maneuvers when 4,500 of Glen Avon's residents pursued their case in a lawsuit against the state, the county, the Stringfellow dump's operator, and dozens of his industrial clients.

In the process of constructing its defense, the state's lawyers requested access to the 125-page interviews conducted with the residents for the health study. The lawyers wanted to search for contradictions between the residents' early responses to the health investigators and their present legal depositions. Despite the pledge of confidentiality in the health study, the judge released the interviews to the state's lawyers.

"Talk about disillusionment," complained Newman. "We had gotten signed promises from the state when we started the study that this stuff would remain confidential. Releasing the information was the ultimate slap in the face."

Neutra insisted that the health department "fought like mad" to block the information's release. "I consider it a very serious impediment to people participating in studies like this in the future," he said.

Newman simply did not believe him. By this time, too much had happened to erode faith and credibility in all authority.

But the strongest indictment of the system that had pitched Newman against Neutra, the citizens of Glen Avon against the State of California, came from the least likely of all sources.

"The state completely whitewashed my report," declared Dr. Dean Baker, the UCLA epidemiologist who had conducted the Glen Avon study. "I gave the State Department of Health Services a five-hundred-page report, some fifty of which contained my explanation of the study's limitations in making conclusions. I ultimately decided that this study was inconclusive. We couldn't tell who in the community was really exposed or not exposed—and the numbers were too small for statistical power. From an epidemiological point of view, this study would *have* to be seen as inconclusive."

Indeed, Baker's criticism of his own study mirrored most of the complaints issued by the community—plus a few they could not have known about.

"The state department of health developed all the public statements about the study," he said. "They did not show them to me. They presented the results of our study as negative—showing no effect. But this study was inconclusive, not negative, and should have been presented to the press and public as being so."

When asked to explain the state's motivations, Baker paused, concentrated, and then spoke very deliberately.

"People in the State health department are well intentioned," he said, "but they are also, after all, government officials. And when they've put a lot of time and money into a study, and must report to the legislature and higher-ups, they have a very difficult time saying, 'we couldn't come to any conclusion'—even though that's truthfully what happened. So they look at the study, see there was no *major* effect, and say there had been no effect at all. Which is not the truth. People in their positions need to say something certain, and '*we don't know*' just doesn't work for them."

Yet Baker also acknowledged that this kind of miscommunication was not simply a matter of the individuals involved; the temptations to blur the close distinctions and subtleties of environmental health studies were overwhelming.

"It's frustrating being an epidemiologist," conceded Baker, "because epidemiology is just not a very effective tool in these kinds of circumstances. There are many situations we *can* deal with. But ultimately, when confronted with the lack of real exposure data in communities like Glen Avon, the scientists who have the strongest backbone will say up front, 'This situation is inherently unstudyable.'"

For Penny Newman, Baker's words sounded less like a vindication than an unexpected echo of the obvious.

"No matter what the study showed," she declared, "we knew there were problems in Glen Avon. So why do the study? We should have put more emphasis on applying political pressure to deal with the commonsense solutions. They needed to stop the exposure, clean up the wastes still up there, and take care of the health problems in this town. It's a political question," she emphasized, "not a scientific question."

- - -

Given the history of Glen Avon, and other towns that have undertaken long and costly health studies, it is easy to see why community activists would conclude that their power to enact change resides in the political realm rather than the world of science. But the citizens of toxic America could not entirely ignore scientific efforts—no matter how frustrating their initial outcomes.

In the long run, citizens and scientists alike had to understand the degree of risk that society had unconsciously assumed by showering the globe with toxic chemicals—if only to calculate how much time, energy, and money should be dedicated to protecting the world from the consequences of their accumulation.

Unfortunately, from the information and analytical tools available today, nobody could state with absolute certainty whether the risk was enormous, nonexistent, or some place in between. Yet one way or another, the nation's political leaders would continue to make decisions about shielding society from toxic harm—even though the wrong decision, erring in either direction, might exact a terrible price.

In terms of money alone, American society faced a perplexing dilemma. If toxics do indeed pose a minimal threat to human health, and can be reasonably assumed to grow no more dangerous as they inevitably pile up over the decades, then the $4 billion spent annually on redeeming a bare fraction of the nation's toxic hot spots, including a smattering of Superfund sites, might well be a pointless waste.

On the other hand, if the health effects turn out to be severe—now, or for future generations—then the trillions of dollars deemed necessary to clean up all the wastes are entirely justified. We can spend the money now; or we can wait and pay later for the diverse economic consequences of chemically induced illnesses—to say nothing of the revenue lost to business and industry because of contaminated lands and waterways.

But given the limitations of the studies conducted in Glen Avon, and other towns throughout the country, there seemed to be no way to make a reasonable, informed decision about society's true toxic toll.

All these problems made it even more surprising when in 1991 the prestigious National Academy of Sciences published a report from its own select Committee on Environmental Epidemiology that boldly stated: "Despite the lack of adequate data ... the committee does find sufficient evidence that hazardous wastes have produced serious health effects in some populations." The committee's report went on to identify "increased rates of birth defects, spontaneous abortion, neurologic impairment, and cancer" occurring in residential neighborhoods undergoing toxic exposures.

How did America's most prominent scientific body—the citadel of procedural orthodoxy—arrive at these conclusions?

The National Academy's Committee on Environmental Epidemiology began its work by reviewing all the health studies conducted in chemically contaminated towns throughout the country—plus all the toxicological data on the chemicals known to be involved. In some cases, the scientists concluded that the wastes had indeed harmed people's health. More often, they admitted that insufficient data rendered a responsible conclusion impossible. But at no time did the committee argue that the lack of a scientific association between contamination and illness meant that the association did not exist. In fact, the scientists warned that the "over-interpretation of epidemioligical results can occur when results that show no effect are believed to prove no effect."

In essence, the committee's report offered the first comprehensive epidemiologic survey of the toxic state of the nation. And once the researchers stepped back from each isolated, individual study to gaze at the larger national picture, they watched a much more alarming problem draw into focus. The scientists combined what they had learned from *all* the studies with what they already knew about a myriad of toxicological mechanisms, as well as information gleaned from animal studies—and only then did they consider the accumulated evidence.

Unexpectedly, this process forced the epidemiologists to discard some of their own rules.

In the course of piecing together the big picture, the researchers did not rely exclusively on the statistical requirement of 95 percent certainty in order to reach conclusions about relationships between toxic exposure and disease. In fact, their own review of the extant health studies convinced them that 95 percent certainty was virtually unattainable. Even more heretically, the scientists suggested that statistics might sometimes be supplanted by "judgment and interpretation."

In other words, if it looked like a toxic illness, swam like a toxic illness, and quacked like a toxic illness—then the mathematical models conventionally preferred to establish its identity as a toxic illness might not be absolutely necessary.

With some pain and self-conscious contortions, the epidemiologists had forced themselves to grasp for "reasonable" conclusions, despite their own professional yearning for certainty. "Public health policy," declared the committee's report, "requires that decisions be made despite the incomplete evidence, with the aim of protecting the public health in the future. . . . Without clear answers, the only prudent course is to err on the side of public safety."

The epidemiological committee's approach to evaluating the health effects of a commonly used industrial solvent, TCE, stood as a good example.

The National Academy of Sciences' Committee on Environmental Epidemiology examined a health study in Tucson, Arizona, in which an unusually high number of heart defects occurred in a neighborhood where Hughes Aircraft and the local airport had contaminated the drinking water with the industrial degreaser, TCE. As usual, the local epidemiologists who studied the community could not link the high rate of birth defects with the toxic exposure. "All this data suggests," said Dr. Stanley Goldberg, who authored the study, "is that we have an area in Tucson in which, if you lived there for a long period of time, the incidence of congenital heart disease is greater than if you didn't."

If no more information had been available, the inconclusive verdict on TCE might have rested there. But Tucson wasn't the only community with water contaminated by TCE that showed high rates of birth defects—and raised a number of epidemiologic questions.

In San Jose, California, the Fairchild Camera and Instrument Company leaked TCA, an industrial solvent similiar to TCE, into one neighborhood's drinking water. The number of babies born with heart defects rose. When the TCA was purged from the water, the rate of heart deformities returned to normal. Once again, following a long study, the state health department concluded that the "solvent leak is an unlikely explanation for this excess" of disease.

Faced with these two studies that could not prove or disprove the association between TCE and heart defects, the Committee on Environmental Epidemiology examined the animal research on the chemical. In 1990, Dr. B. V. Dawson had contaminated the wombs of pregnant rats with TCE. The offspring exhibited high rates of heart defects. The same experiment was conducted with chickens—reaping the same deformities. But of course, rats and chickens are not simply miniature human beings. By definition, scientists had to regard the animal studies as inconclusive; on their own, they did not prove that TCE causes heart defects in humans.

San Jose, California. Bari and Dustin Kitchens at well thirteen, the major source of TCA-contaminated water. Dustin was one of 117 children with birth defects, cancer, or blood diseases who reached a financial settlement in a lawsuit against the company responsible for the toxic leak.

Yet when the committee drew together *all* the available evidence regarding the relationship between TCE and heart defects in babies, its members had no trouble reaching a conclusion. "Several lines of evidence," their report stated, "point to a causal nexus between exposure to TCE and cardiac congenital anomalies." While no individual study had shown with scientific certainty that human heart defects were produced by exposure to the solvent, the committee had nevertheless concluded that the *weight* of evidence did point to a cause-and-effect relationship.

"If we waited for 100 percent prediction," said Dr. Bernard Goldstein, who chaired the committee, "we would not be protecting the public."

In fact, the chairman's perspective exactly mirrored the point of view expressed by the chief administrator of the Superfund program.

"There are forty million people out there who live within four miles of a Superfund site," said the program's director, Richard J. Guim. "Do you tell them, 'Hey, sit tight, we're studying this problem? Until we have better information in twenty years, we won't take action'?"

This method of grafting common logic on to scientific scrutiny had enormous implications for public policy decisions—and they reached beyond the redemption of Superfund or other residential toxic hot spots. As the law now stands, individual chemicals, like individual citizens, remain innocent until proven guilty. The government will not curtail a chemical's use until its harm has been scientifically established. The National Academy of Sciences' Committee on Environmental Epidemiology seemed to imply that the burden of proof, in some instances, should shift to the suspect.

Oddly enough, this is exactly what happened in one case involving TCE. In 1989, consumer groups sued the manufacturers of Liquid Paper to compel them to stop using the solvent in their typewriter correction fluid. Despite the lack of conclusive scientific evidence that the chemical was harmful, the consumer groups managed through legal pressure to persuade the manufacturers that the continued presence of TCE fumes at practically every desk and workstation in the United States might present a health problem—or at least the threat of corporate liability and endless litigation. The company acquiesced, and also put out a new product without the TCE, ironically named Mistake Out. And while the company complained that its new correction fluid was "slow drying and it smudges more," few people found their lives significantly diminished by these imperfections.

To Dr. Elizabeth Whelan, president of the American Council of Science and Health, this kind of turmoil and change in the marketplace indicated precisely what was wrong with the methods used by the National Academy of Sciences' Committee on Environmental Epidemiology.

"Decisions about chemicals should be based on hard, peer-reviewed science," Dr. Whelan told us emphatically, "not intuition, politics, or emotion."

Throughout her career, Whelan's extensive research and writing had rightfully extolled the innumerable human benefits reaped from the chemical revolution: improved agricultural harvests, synthetic antibiotics, and countless labor, time, and cost-saving products that most of us now take for granted. She didn't want any of these achievments jeopardized by misguided reformers. And she took particular objection to the Committee on Environmental Epidemiology's method of combining evidence from numerous studies to arrive at conclusions about the health effects of hazardous chemicals.

"I hate that approach," she said. "It's like saying, 'The individual studies didn't show anything, but if I put them all in a stew and stir them up, maybe we can get something out of it.' If some health effects were really significant," she argued, "it would show up in any study. Study after study would show the effect."

In an ironic affirmation of the views expressed by Penny Newman and other community activists, Whelan insisted that the health studies were, in fact, "politically" motivated.

"Superfund is coming up for renewal this year," she observed, "and they're desperately seeking justification for its existence. Finding diseases caused by the sites would fit very nicely for them. We should put our resources into the known causes of disease, and pay no attention whatsoever to these purely hypothetical things until we solve the other causes. This inferential approach to science is not only detrimental to our whole public health system, it also threatens our standard of living. If we assume that even the most minute chemical exposures to humans are intolerable, we'd have to ban everything."

In her book, *Toxic Terrorism*, Dr. Whelan took this line of thinking to its furthest end: "We might have to return to the traditional seventy-hour work week, working at least six days, weaning ourselves from the labor-saving devices that made life easier. . . . We would have to do with less food, fuel, clothing, and other goods. . . . I am incensed by the possibility that lies, exaggeration, and innuendo about the relationship of our health and environment might change the whole fabric of the lives of Americans."

Whelan's criticism would probably seem more compelling if the scientific studies meant to protect citizens from toxic harm appeared to be functioning at top form. But even the federal government harshly criticized the performance of its own scientific regulatory agencies. According to a General Accounting Office survey, the health risk studies conducted at nearly one thousand Superfund priority sites were "seriously deficient" and "generally have not been useful." And as Whelan herself concluded, more than a thread of politics ran through most epidemiological investigations. The scientists and authorities, it seemed, were vulnerable to the vociferous criticism from all sides, practically all of the time.

During our stay in Glen Avon, we came to learn how frustrating, and ultimately counterproductive, the contention between the scientists and the community could be. But the problem did not seem to reside with the people, on either side, involved in trying to unravel Glen Avon's mystery. In fact, it would be hard to find better-intentioned individuals than Penny Newman, Shirley Ann Neal, Dr. Dean Baker, and Dr. Raymond Neutra.

The real trouble lay in the limitations of the science—and the inevitable tendency of governmental bureaucracies to protect their own interests by obscuring uncomfortable subtleties, and then drawing broad conclusions from maddeningly inconclusive findings.

"I clearly understand now how society makes its decisions," said Penny Newman. "I used to think it was based on facts and reasoning. I'm much more skeptical now of things that are presented to me as 'facts.' In Glen Avon, we've begun to trust our instincts more."

In Glen Avon, and in contaminated communities throughout the country, the breach between the scientists and the citizens seemed to be widening. In the blunt words of Penny Newman: "I've learned to trust what the community feels, and not what some jerk at the health department says. We found that often what the community said was right, and what the experts said was wrong."

Yet if Glen Avon proved anything, it was that the citizens and the scientists needed one another more than ever. The terms of their relationship might still require a great deal of work, but the insistent demand that crucial decisions regarding toxic chemicals be made by citizens, and informed by science, was not going to fade away.

Love's Uncertain Chemistry
The Tragedy and Romance of Environmental Illness in the West Texas Desert

Texas. The Davis Mountains.

We had just left the Southern California desert community of Glen Avon, where an entire town had been exposed to toxic chemicals—and as a result, thousands of residents claimed illness from the encounter. Our next destination was a much larger and more imposing desert: West Texas, one of the most remote and unblemished regions of the country. We were heading there to meet some people who composed a very different kind of community, constellated around an immense fear of chemicals that far exceeded the concerns of anybody we had met so far.

In Glen Avon, we had learned that the community health effects of toxic exposures, at any level, were never clear-cut. Not in Glen

Avon, not in Love Canal, not in Times Beach, not in any town we had visited; not anywhere. But scientific uncertainty had contained its own truths; it simply demanded patience to unravel them. Fear, we would soon discover, also had some important lessons to teach.

In fact, nowhere would we encounter more instructive fears than in the unlikely communities of West Texas, among the toxic refugees who had isolated themselves from the modern world.

- - -

In 1985, Robert McIntyre fled the modern world to take his stand in the West Texas desert.

He loaded up his old Chevy truck with supplies and flew out of Austin, driving across the vast and empty Edwards Plateau to ascend the sawtooth ridge and high mesas of the Trans-Pecos—where Comanches once slaughtered settlers, black American "buffalo soldiers" beat back Apaches, and the Confederate army base camp of Fort Davis (named for the South's commander-in-chief, Jefferson Davis) fought Union troops to a bloody draw for nearly one year. Into this contentious landscape, Robert McIntyre came searching for peace, quiet, and clean living. He headed straight up the Davis Mountains, following a narrow dirt path past tangles of mesquite and manzanita until arriving at a six-thousand-foot mountaintop sanctuary called High Lonesome. On this isolated peak, Robert McIntyre took up residence in a one-room shack without electricity or telephone. Each morning, he peered down at the sagebrush desert unfurled like a flat, scrubby carpet all the way to Mexico. The air he breathed was so clear and clean that astronomers at the McDonald Observatory sited one of the world's most powerful telescopes nearby to scan the skies for satellites, stars, pulsars, and quasars glinting millions of miles away.

Robert McIntyre had never seen such gorgeous and pristine countryside. But the 54-year-old former oil finance lawyer had not moved to West Texas for the scenery. He believed that if he lived any place in the United States other than this barren perch on High Lonesome, chemical pollution would kill him.

McIntyre claimed to suffer from a disease called "multiple chemical sensitivity." People with similar problems referred to this mysterious ailment as "environmental illness," "allergy to the twentieth century," or simply "twentieth-century disease." Oftentimes the people who identify themselves as chemically sensitive display painful, inexplicable, and often highly dramatic symptoms. Some swoon into a sudden fit of muscle spasms when they catch a whiff of perfume from a passing stranger. Others grow nauseated from exposure to a

Robert McIntyre on the porch of his trailer in the Davis mountains.

newly painted room in their own house, and remain debilitated for weeks, months, or even years. Chronic sufferers endure blinding headaches, vomiting, severe abdominal pain, temporary paralysis, and a descent into a lingering, low-grade listlessness and loss of well-being, like a middling case of the flu that never really goes away.

Yet for every person who complains of environmental illness, and for any doctor who claims to diagnose and treat it, several experts stand in line waiting to argue that the disease doesn't even exist. According to many physicians, and virtually all psychiatrists who have investigated the subject, environmental illness remains a disease of the mind, not the body. The chemically sensitive and their advocates respond that their symptoms indicate an entirely new malady, spawned by the unprecedented profusion of toxic chemicals through-out the world. They insist that its complexities demand new ways of thinking about health and the human body. To this end, they use the analogy of a rain barrel to explain the disease. A rain barrel may hold 30 gallons of water. But once it fills to the top, every additional drop of rain will flow over the edge. The human immune system, they sug-gest, works in a similar fashion: It can only contain so much contam-ination before overflowing into illness. According to the theory, indidividual immune systems, like rain barrels, also vary in container capacity. While minute amounts of chemicals may cause no symptoms in one person, another person whose immune system is already filled to the brim by previous exposures could suffer severe reactions. For chemically sensitive people, whose rain barrels are completely topped off—and possibly damaged from traumatic overloads—even immeasurably small quantities of seemingly safe, everyday chemicals can produce hideous symptoms.

The rain barrel analogy is neat, clean, and logical. But its advocates have yet to prove in laboratory or clinical settings that the human immune system functions in this way. When Robert McIntyre abandoned the modern world for his home at the top of High Lonesome, he had already heard countless lectures from conventional doctors who doubted the very existence of the disease he believed was destroying him. But as chemicals engulfed his world, he grew convinced that he had no choice other than to leave.

McIntyre's symptoms began in law school. Whenever he opened a textbook, his head pounded and the muscles throughout his taut, athletic body throbbed and ached. He now believes that chemical fumes wafted up from the inked pages of his law books and attacked his immune system, which had been previously damaged by some severe, but unrecognized toxic exposure. Today McIntyre grows nauseated if he comes in contact with detergent on anyone's clothing. The scent of after-shave or deodorant doubles him up with stomach cramps. When traces of agricultural pesticides float through the air in quantities imperceptible to other people, McIntyre is driven to his bed for days. Years of successive exposures to the synthetic substances of everyday life have worn him down to a state of acute petrochemical vulnerability, making him unfit for the modern world.

McIntyre felt safe for the first time in years in the seclusion of High Lonesome. Atop the Davis Mountains, he quarantined himself from the rest of the world—not because of any illness that he might spread, but due to his intense sensitivities. Over the past few years, some 30 other environmentally ill people also have collected in this part of West Texas. They arrive to set up their own bunkers amid the howling grey wolves, wild boars, and javelinas; they build toxin-free homes in sheltered bends, and retreat to their own mountain hermitages.

One afternoon, Robert spent 45 minutes hiking down High Lonesome's steep deer path to meet his closest neighbor, an environmentally ill woman named Kari Pratt. Kari lived with her two children in a small trailer nestled into a grassy mountain cove. When Robert first arrived, she refused to emerge from the trailer's kitchen, explaining that she had covered her hands and arms with peanut butter.

"I had no idea why," confessed McIntyre. "But one of the things I know about environmentally ill people is that they do things differently. Whatever they're doing is for their own best interests. I don't question it."

Robert's low, soothing drawl eventually put her at ease. Kari washed the peanut butter off her hands and arms and stepped outside to greet him. She was a pretty, reed-slender, 43-year-old woman with deep-set, tragic eyes and the haunting smile of a frightened child. "Her hands," Robert recalled fondly, "were as white as snow. She had rubbed and scrubbed and cleaned them so hard."

Their first meeting lasted only a few minutes. They talked about their lives and illnesses. Kari shyly explained about the peanut butter. "If you have grease, oil, or chemicals on your skin," she told Robert, "it all gets drawn into the peanut butter." Robert nodded agreement. Over time, he tried it himself. "Peanut butter is fantastic," he confirmed. "It works on metal, plastic, anything. When our car is at the shop, they use a chemical solvent that makes us ill. So we just peanut butter the engine and leave it there for a week. When we wash off the peanut butter, we can tolerate being near the car again."

From the beginning, Kari felt curiously drawn to Robert. "He had a very warm and caring smile," she said, "and he seemed so kind, and to really care about my kids." But the sudden attention from this genial and ruggedly handsome man, out in the middle of nowhere, also made her uneasy. "When we first met, I was so embarrassed," said Kari. "My hair wasn't even combed. And I just wondered, 'What does he want of me?'"

Eventually, Kari would think of Robert as the answer to her prayers.

Kari Pratt had left her home and husband in Des Moines, Iowa, after enduring years of mysterious muscle and joint pains, fatigue, and severe depression. Kari believed her problems stemmed from a chronic uterine infection caused by the infamous Dalkon shield birth control device. The infection, she asserted, had damaged her immune system so that any exposure to synthetic chemicals could produce terrible physical and emotional pain. Conventional medical treatments only made Kari's ailments worse. In a desperate attempt to avoid all chemical exposures, she absconded to the Davis Mountains and spent most of her days and nights outdoors, basking in the desert's clean, clear air. She knew that she could not run any further. If the Texas desert did not provide relief, she would commit suicide.

"I marked the spots on my wrists where I could make the cuts," she said. Then she sat in the desert and prayed.

"God, if you are real and you're a God of love, send some love into my life. Show me your love. I just can't handle life."

Kari's first meeting with Robert offered some hope of emotional comfort. But following their conversation, her pain and fatigue increased, immobilizing her for two full days. Robert also enjoyed their talk, but as soon as he got back to High Lonesome, severe headaches gripped him once again. For months, his health had been improving. This sudden, unexpected attack dismayed him.

The following week, after an argument with her 12-year-old son, Kari climbed the steep trail to High Lonesome and sought Robert's counsel. He missed his own children back home, and he seemed a natural with kids. His advice reassured Kari. "It was such a relief to have somebody here to help."

After their second encounter, Kari and Robert realized that they had to contend with a terrible, almost unbelievable problem. "This is going to sound like 'The Twilight Zone,'" admitted Kari, "but we would enjoy being with each other, and then we'd go home and be sick for days. Robert would have his migraine headaches. We made each other ill." According to the couple's self-diagnosis, the chemicals in each another's bodies precipitated the awful response. They were "allergic" to each other.

But they wouldn't give up. Kari and Robert experimented with various ways to safely spend time together. They sat outside in the clear mountain air and talked for two or three hours at a stretch— always separated by a distance of 15 feet. Over six months, they followed an exacting regimen, both eating the same foods, taking the same medications, using the same skin lotions. Their reactions diminished. They moved closer, talking at a distance of nine feet. Then five feet. Then three feet.

After months of diligence, remixing their body chemistry to as near a perfect match as they could manage, they were finally able to draw near enough for an intimate embrace. "Isn't that incredible?" asked Kari.

- - -

We had traveled more than fifteen hundred miles over ten days, zigzagging across Texas to visit the nation's most remote outposts of environmental illness. Among the dozens of people we encountered, Robert McIntyre and Kari Pratt proved striking, although not singular examples of the idiosyncratic behavior among chemically hypersensitive people that had earned them a controversial reputation. In fact, we were already somewhat familiar with environmental illness in its less dramatic presentations. In virtually all the contaminated communities we had visited, from Yukon, Pennsylvania, to Casmalia, California, there were always several people who identified themselves as environmentally ill. In order to understand their predicament, and more importantly, to grasp the larger story of chemical contamination throughout the nation, we needed to explore the syndrome's furthest shores. In the secluded small town of Wimberly, amid the rolling hills of East Texas, we met Dot Dimitri. The 72-year-old widow had just started to recount her own story of flight from the petrochemical world when she abruptly halted, looked grievously alarmed, and asked us to leave her home. One of our sweaters, she explained, was making her ill. This sudden sensitivity to the synthetic chemical dyes of the bright coral-green sweater did not surprise us. Many environmentally ill people complain about "out-gassing"—the natural process of chemicals rising off new clothing or other treated materials, like steam rising from a kettle of boiling water, as the molecules change from a liquid or solid into a gas.

Anybody who has walked into a recently painted room or commented on "new car smell" is familiar with out-gassing. Most people shrug off these everyday chemical odors. Others object to the nauseated feeling they get from lingering in a freshly painted room.

At the furthest end of this spectrum stand people like Dot Dimitri, who say that they cannot spend even a few minutes near a sweater manufactured many years ago with synthetic dyes. The minute amounts of chemicals that continued to out-gas from the sweater, not noticable to either of us, were enough to overwhelm Dimitri and make her ill.

To continue the interview, we settled on a compromise with Dimitri. We stepped outside and sat on two chairs positioned on her front porch, while she remained inside at a distance of several feet. Then we asked our questions through an open window, as though ordering burgers at a Jack-In-The-Box drive-up window. Another half hour into the conversation, Dimitri casually mentioned that electromagnetic waves broadcast from our skulls enabled her to read our thoughts. We ended the interview and moved on.

Driving away from Dimitri's home, we felt foolish for having traveled halfway across the country to talk with a mind reader and two hermits who covered their limbs with peanut butter. But we could not simply dismiss Dimitri, McIntyre, and Pratt as cranks whose plight bore no relevance to our nation's mounting contamination. Although many of the environmentally ill we encountered appeared eccentric

One of the authors interviewed Dot Dimitri from her porch while she sat inside, protected from the chemicals on his sweater. Dimitri also hung her family photographs on the outside of the house.

in their behavior, they still reflected society's widespread anxieties about the unpredictable effects of toxic chemicals on human health. And there seemed no way to immediately ascertain that they weren't absolutely correct—that for some of them, their claims that trace amounts of chemicals indeed made them sick.

Nobody knows for certain how many people suffer from environmental illness. In 1989, the National Foundation for the Chemically Hypersensitive conducted interviews with 6,800 people who claimed to have the disease. The National Academy of Sciences approximates that 15 percent of the population may experience "increased allergic sensitivity" to chemicals. In any case, the ranks of the environmentally ill are growing, with support groups popping up throughout the country, contributing to the most austere vision of what life in a contaminated world could mean.

Beyond their capacity to affect the popular imagination, another important reason existed for seriously regarding the claims of the environmentally ill. Dot Dimitri's mind reading and Kari Pratt's peanut butter might *seem* crazy, but the imperfections of the messengers did not necessarily invalidate their message. If we assumed for the moment that the two women did suffer from some disease whose cause eluded comprehension at this time, then their behavior, however odd, seemed justified.

Madness, neurosis, or everyday eccentricity are always defined by norms of the day. But disease tends to warp the norm, sometimes beyond recognition. To our eyes today, the tubercular patient of the 1800s would seem a very odd creature. Prior to the discovery of the bacillus that causes tuberculosis, tens of thousands of "consumptives" flocked to places as isolated as the Davis Mountains of West Texas, hoping in vain that fresh air and clean living would heal them. "The TB sufferer was a dropout, a wanderer in endless search of the healthy place," wrote Susan Sontag in *Illness As Metaphor*. " . . . Starting in the early nineteenth century, TB became a new reason for exile." Inside the sanatoriums, countless reasonable young women—perhaps some resembling Kari Pratt—gradually descended into sickness, losing weight and energy, lapsing into debilitating fevers, fits of coughing, and complaints of constant pain. Their journies to the sanatorium seemed a last resort. After a woman endured months of ineffectual treatments, when her husband and children finally visited, they would encounter a very different woman than the wife and mother they had once known. At lunch, the patient might pick through her food, lecturing about the restorative properties of mud packs and hot sassafrass tea. Perhaps she would conclude that a diet of carrots would give her the best chance to survive. Since cleanliness was the hallmark of most sanatorium regimes, she might depart from the table every ten minutes to compulsively wash her hands.

All this sounds like classically neurotic behavior, or worse. But given our perspective on the limitations of nineteenth-century medicine, we forgive the distraught patient of the past. A psychiatrist of today visiting a nineteenth-century sanatorium would almost certainly conclude that the patients exhibited symptoms "characteristic of several well-known psychiatric disorders"—as one contemporary psychiatrist recently characterized patients he evaluated who claimed they suffered from chemical sensitivity. But today's psychiatrists would also understand that the tubercular patients' "crazy" behavior resulted from the stress of the misunderstood illness.

Today, if we accept the possibility that some form of environmental illness may exist—even if it has nothing to do with the unproven rain barrel theory—then our presumptions about the mental health of its sufferers no longer stand cozy and pat. In the face of a life-threatening illness that the most learned specialists cannot tame or comprehend, desperate behaviors will thrive amid boundless anxieties.

Yet today's psychiatric establishment has assumed a stance of uniform hostility towards environmental illness, refusing to concede that sick people, by definition, do not behave like healthy people.

Doctors in Toronto, Canada, conducted a study in 1985 with 18 environmentally ill patients, and found that "all of them were suffering from a recognizable psychiatric disorder." Yet 16 of the 18 patients had refused to undergo psychological testing. Although each patient agreed to an interview, "10 did so reluctantly." The psychiatrists judged the patients' decision to "vigorously resist psychiatric referral" as verification of their mental illness, concluding from limited observation that virtually all the patients' symptoms were "characteristic of several well-known disorders."

Other studies have employed similar methodologies in which the patients' reluctance to admit that they may have emotional problems is taken as proof that their physiological disease does not exist. Dr. Carroll Brodsky, a professor of psychiatry at the University of California at San Francisco, and one of the most influential researchers in the field, has stated that most of the environmentally ill people he has studied were " . . . acutely aware of their bodily functions and believed that their symptoms were physical rather than mental, despite contrary opinions from psychiatrists and other physicians." In other words, Brodsky's patients established their mental instability by refusing to admit to it. Writing in the *Journal of the American Medical Association*, Dr. Donald Black, a psychiatrist at the University of Iowa, similarly maintained that patients referred to him with complaints of environmental illness demonstrated a higher rate of psychiatric illness than the "normal" population. Yet Black noted that he began his research with "the premise that the syndrome was not a true medical disease." Of the 26 patients interviewed, most appeared to suffer from depression, a psychological state common to patients with both real and imagined illnesses. "It is my belief," Black concluded, "that people diagnosed as having environmental illness in most cases do have something wrong: a garden variety emotional disorder."

To psychiatrists like Black, Brodsky, and a handful of others who have investigated the psychological implications of environmental illness, the peanut butter treatments and 15-foot conversations of Kari Pratt and Robert McIntyre must certainly appear abnormal. But the psychiatrists' pronouncements of mental illness only make sense because they assume the physical disease does not exist. In several cases, the studies debunking environmental illness have even been funded by insurance companies, who maintain a strong interest in keeping the uncertain malady off their roster of actionable claims—thus further eroding the appearance of objectivity among the researchers. When a cancer patient deserts all orthodox treatments and resorts to megavitamin therapy, or travels to Mexico for Laetrile treatments, the patient's response to the disease might be considered "abnormal," but nobody concludes that the patient is not physically ill. With environmental illness, the psychiatrists reason in a circle: The disease does not exist; people who believe that they have the disease must be emotionally disturbed; if they are emotionally disturbed then their disease is psychological, not physical; therefore, the disease does not exist.

Today many environmentally ill people not only assert the reality of their disease, but they consider their plight a harbinger of future public health catastrophes. In that well-worn phrase, the environmentally ill now assert *their* claim as the true canaries in the mine shaft. What Dimitri, Pratt, and McIntyre feel today, the rest of us will suffer tomorrow. Of course, all these arguments do not allay the fact that mental instability, or even the collected weight of life's everyday burdens, can and often does lead to imagined illness. Despite a tendency to slant their studies with preconceptions, the psychiatrists could be right. Environmental illness might simply signify the latest incarnation of an age-old, self-selected aristocracy of the ill. "I should like to die of consumption," Lord Byron confessed to a friend. ". . . Because the ladies would all say, 'Look at that poor Byron, how interesting he looks in dying.' " Hypersensitivity to petrochemicals may be unconsciously regarded by its sufferers as a similar mark of distinction. Even if environmental illness does not exist as a genuine physical disease, it certainly provides an extraordinary opportunity for escape; or, as Sontag described one renegade aspect of TB—"a way of retiring from the world without having to take responsibility for the decision. . . ."

"Sometimes I think being ill and dying aren't serious at all," declared a character convalescing from TB in Thomas Mann's *The Magic Mountain*, "just a sort of loafing about and wasting time up here; life is only serious down below."

-    -   -

"It was really a happy time," recalled Kari Pratt, as she sat upon a grassy knoll in the Davis Mountains, gazed out over the West Texas desert, and talked about her life in Des Moines before she became ill. Pratt's first child was born in 1968, with a second and third to follow over the next five years. "It was so much fun to have little kids. I did lots of sewing; I even supervised a hobby class for our church."

The Pratt family's entire week revolved around their church. They attended services on Sunday mornings and evenings, as well as on Tuesday, Thursday, and Friday nights. Members of the congregation adhered strictly to the church leaders' teachings. Women could not cut their hair. They remained silent in church. The church rules forbid children from playing sports on school teams.

As the Pratt children grew older, the happy times faded. When Kari complained to her husband that the church's severity squeezed the joy out of life, he told her, "You have a very rebellious mind." After the birth of her third child, Kari was fitted with a Dalkon shield IUD. Almost immediately, terrible burning pains shot through her abdomen, head, hands, and fingers. "I just burned all over," she recalled with a shudder. "It was like somebody was cutting me with knives."

Her pain would flare particularly on days when the neighbors sprayed their lawn with an unpleasant-smelling fungicide. When her gas stove began to bother her, she removed it from her house. One day, the family was driving home from a funeral when they got stuck on the road behind a large diesel truck. The diesel smell made Pratt so sick that she took to bed for three days.

About this time, Pratt's mother heard a radio talk show broadcast from Des Moines that featured Dr. Theron Randolph, the physician who in 1951 coined the term, "environmental illness." She telephoned Kari. "Turn on the radio," she ordered, "and listen to this." Pratt sat down and listened hard, alternately anxious and hopeful about what she heard. She urged her husband to call the radio station and relate her case to Dr. Randolph. When Pratt's husband told the doctor about the diesel truck, Randolph flatly explained: She is reacting to chemicals, and needs to be in the hospital immediately.

Randolph put Pratt in the hospital. At first, she expected an immediate cure. She only grew worse. "I just reacted to everything. And back then, every time you had a reaction, they'd give you an enema." Pratt's five-foot, four-inch frame dwindled from 131 to 93 pounds. She couldn't move, she couldn't talk. The slightest noise hurt her ears. "I was so sensitive that I couldn't even stand a fan in my room." A nurse stuffed Pratt's ears with cotton. When that didn't work, the nurse wrapped her head in a gauze turban. "I must have looked like a person from outer space!" said Pratt. "People could see into the room. I wondered, 'What do they think is going on in this hospital?'"

- - -

When Dr. Theron Randolph proposed his theories of "food allergies" in the 1950s, he immediately drew fire from his fellow specialists.

The problem involved a crucial semantic distinction: the definition of the word "allergy."

As a Harvard-trained, board-certified allergist, Randolph knew that the medical establishment used the term in a very precise and narrow way. In modern medicine, an allergic reaction involves the immune system's production of a protein called Immunoglobulin E (or more commonly, IgE). The body manufactures IgE protein in response to known allergens, such as pollen or ragweed. The presence of this protein causes the familiar allergic reactions, such as a runny nose, itchy and watery eyes, asthma, mild skin rashes and hives—or in the case of a bee sting, the potentially fatal anaphylactic shock. All these allergic responses function through biologic mechanisms mediated by IgE and its attendant body chemicals in processes charted and understood for many years by the field's specialists.

Randolph claimed that allergies involving a great number of common foods caused a vast range of chronic illnesses. Wheat and dairy products, for instance, might cause migraine headaches or arthritis. But he offered no evidence that IgE played any part in the process. Conventional allergists thought they had demolished Randolph's theories when they tested people with these illnesses and found they had normal levels of IgE. But Randolph persisted, expanding his theory to claim that for some particularly sensitive people, small amounts of chemicals could cause an even wider variety of diseases. Randolph also detailed the lesser symptoms of chemical allergy, which included a very long list of common ailments, largely of a subjective nature, such as fatigue, profuse sweating, headaches, itching, depression, anxiety, weight changes, and bad breath.

Randolph predicated his theories on the concept of "total load"—the idea that illness results from a combination of stresses, including adverse reactions to foods and chemicals, as well as other more generally accepted compromises of the body and mind, such as infection and psychological trauma. Most insidious among these new stressors were the chemical contacts of everyday living: deodorants, perfumes, hairspray, car exhaust, bleaches, dyes, chlorinated and fluoridated drinking water, mouthwash, shampoo, and even felt-tipped pens. Randolph's list ran on and on. He also theorized that the stresses promoting illness varied among individuals. In other words, everybody is born with a different-sized rain barrel that can sustain to varying degrees the wear and tear of persistent chemical exposure. Randolph believed that fluctuations in the dose and frequency of these assaults prevented most clinicians from grasping the role of chemicals in causing illness.

Allergists rose in a wave to condemn Randolph's theories. Research failed to show that IgE, and thus allergic reactions, played any part in chemical sensitivity, let alone spawned the wide range of diseases that Randolph claimed. As his public notoriety grew, Randolph's colleagues launched their counterassault. Administrators at Northwestern University Medical School curtailed his teaching duties, citing his "pernicious influence on medical students." In the allergists' view, Randolph ranked as the era's most prominent snake oil salesman.

Long before this point, Randolph and his critics might have benefitted from a simple consultation—not with a board of scientific experts, but rather with a linguist. Conventional practitioners defined "allergy" exclusively in terms of IgE response. Randolph used the language of everyday speech, in which nonphysicians referred to many things that bothered them physically as an "allergy." If a person disliked warm milk to the point of queasiness, he might claim a "milk allergy." If a medicine made a patient nauseated, he would inform his doctor that he was allergic to it—even though a simple chemical irritation of the stomach lining probably caused the reaction. Randolph's patients who experienced unpleasant responses to small amounts of chemicals indeed were *sensitive* to those chemicals. But when Randolph expropriated the word "allergy," his former colleagues blared their objections.

In 1965, Randolph joined forces with a number of like-minded physicians to found the Society for Clinical Ecology. Although not recognized as a genuine medical specialty in the United States, the ranks of clinical ecology have grown steadily over the past four decades. In 1989, over five hundred physicians in various sanctioned medical specialties had trained in clinical ecology. In the words of one of Randolph's disciples, clinical ecology was ideally suited for "patients who do not get well under the present system of medical practice. It offers them and their doctors a different way of looking at ill health." The problem remains that for most conventional researchers and clinicians, the theories of clinical ecology work best on a metaphorical, not scientific level.

"The theory of clinical ecology," summarized one of its adherents, Dr. Richard Mackarness, "is that most human illness is a result of breakdown of interrelated parts of the body due to inability to remain normal in the face of daily exposures to specific chemicals put into our food and drink and into the air we breathe. Under modern conditions of chemical pollution in the industrialized world, the theory continues, our powers of adaptation designed for Stone-Age conditions are running out. The more susceptible among us are becoming chronically sick.''

This vision certainly has its appeal, especially in a society growing evermore aware of environmental problems. The plight of environmentally ill people seems to mirror the condition of the entire world today: Both sustain daily exposures to a vast number of synthetic chemicals, with an infinite number of combinations and doses. Nobody can really say what these exposures mean in terms of human health.

Yet the clinical ecologists claim that they *do* know. They view themselves as part of the long line of scientists and visionaries persecuted in their time, vindicated in the future. Today, most clinical ecologists admit that no evidence exists to indict IgE in the mechanism of chemical sensitivity. Yet the traditional allergists still assail the clinical ecologists on this count, while virtually ignoring the possibility that some other mechanism of "sensitivity" may exist. The allergists have decided that the disease of chemical sensitivity is not real, based on the clinical ecologists' incorrect usage of the term "allergy." This leap in faith and logic seems the equivalent of orthopedic specialists objecting to the theories and practices of chiropractors, and thus concluding that back pain doesn't exist.

Doctors Nicholas Ashford and Claudia Miller are scientists who have looked into the issue of multiple chemical sensitivity in great depth. In their 1991 survey executed for the New Jersey Department of Health, they noted that allergists are frequently called on to provide expert advice about multiple chemical sensitivity, when it may very well be that the problem has nothing to do with allergy but is a non-allergic toxicity.

"Allergists continue to point to the scientific basis of their practice and their detailed knowledge of immune mechanisms," concluded Ashford and Miller. "Clinical ecologists stress the importance of their clinical observations. To some degree, their conflicts are an extension of the traditional tension between academicians and clinicians—a tension that has served neither side well."

Stuck in the middle of the battle between clinical ecologists and allergists are the long-suffering patients, held hostage to an ever-widening, hostile debate.

- - -

Kari Pratt returned to her family after Randolph released her from the hospital. But her illness persisted. Randolph referred her to Dr. William Rea, founder of the Environmental Health Center in Dallas, Texas. Rea's hospital unit offered complete isolation from synthetic chemicals. "I was one of the first ones in that new unit," said Kari. "It got me on my feet again, but it was really lonely and scary. I was all by myself, I didn't have my family, I didn't have my husband. I wanted so badly to get well."

After treatment, Pratt returned home to Iowa. But when her husband picked her up at the airport, she immediately sensed something wrong. In the car, as Pratt now recalls, her husband explained that the brothers of their church had looked into her illness, talked with allergists and other experts, and concluded that her problems lie in the psychological realm. They wanted to commit her to a mental hospital.

Kari opened the passenger's door and jumped out of the moving automobile. She stumbled down the sidewalk, located a telephone booth near a used car lot, and called the police. The police urged the Pratts to go home.

When Kari finally returned home, she found a church brother waiting for her. After a half-hour harangue, she finally agreed, "for the church's sake," to visit a psychiatrist. "It was just supposed to be a consultation," she said. "But once they got me there, they called a judge and had me committed."

In the hospital's locked ward, Kari covered her face with a cloth to protect herself from chemical exposures. The other patients smoked cigarettes, the carpets out-gassed horribly, the staff scrubbed their clothes in detergent. "When are you going to come out from behind that mask?" asked the other patients. Kari thought only about escaping, but she felt too weak to even put on her clothes. She wore her nightgown for days, until she learned that when "the crazies got dressed," the psychiatrists presume they are getting better.

"I put on my best clothes," she said. "As sick as I was, I started to act very positive. I told them their treatment was helping me. Then I did what you have to do when you're fighting for your life. I promised my husband sex. I promised to help him take care of the kids. And I prayed every minute. Every breath was a prayer to get out of there. It was hell on earth." When the hospital finally released Kari, the brothers of the family's church convinced her to stop taking the medicines prescribed by the clinical ecologists. "I thought, maybe they're right. They say they have the mind of God." But her pain and illness grew so bad that once again she could barely move.

When a sympathetic older woman from the church telephoned the Pratt home, Kari told her the entire story. "She had a fit," Kari remembered. "She said that if the church brothers ever came to my house again, I should tell them to get out." Kari resumed taking her medicines. One year later, in June 1984, at the suggestion of Dr. Rea, medical director of the Environmental Health Center in Dallas, Kari left Des Moines and moved to the West Texas desert.

- - -

"I'm the resident quack," declared Dr. Gerald Ross, one of the chief physicians at the Environmental Health Center. He laughed off the label with a shrug of resignation as the turbine whoosh of the clinic's huge air-filtering system nearly drowned out his voice. Enormous pipes ran along the ceiling to suck out every conceivable contaminant and refill the room with clean, cool, thoroughly filtered air. Thin blue porcelain and steel plates covered the walls to prevent chemicals from leaking into the rooms from construction materials. Special carpets guaranteed not to out-gas trailed down the hallways.

Dr. Gerald Ross is a slender, rather shy, boyishly handsome, 42-year-old general practitioner who turned to clinical ecology when his own health failed. After medical school, he began a busy family practice in a small town in Nova Scotia, where he handled everything from breach births to heart attacks. When Ross caught a viral infection, similiar to mononucleosis, he took some time off from his practice, expecting to regain his full strength over a few weeks. But he never did. His symptoms lingered for years. His muscles hurt, his head pounded with ferocious headaches, his heart beat arrhythmically, and his entire body felt ground down by an "absolutely profound fatigue." Even worse were "the burning, scalding, uncontrollable pains" that cut through his body like the knives that had tormented Kari Pratt.

"I just felt so unwell," admitted Ross, in a feat of Canadian understatement. When a legion of specialists failed to identify the problem, Ross grew discouraged and demoralized. He consulted a psychiatrist who had been one of his professors in medical school. The older man insisted Ross' feelings seemed an appropriate response to the inexplicable health problems that had undermined his career and plunged him into near-total disability. "He told me I was more fed up than depressed. There was nothing he could do."

Finally, Ross underwent tests at the Mayo Clinic, which revealed nerve damage throughout his body. "The actual degeneration of the inside wiring of the nerves finally explained the scalding sensations on the surface of my skin," said Ross. But the degeneration's cause remained murky. Soon after the tests, Ross noticed a pattern to his pain. "When I ate certain foods, my muscles ached and the burning skin would get worse. If I drove behind a diesel truck, I'd have to drop way back. My joints would pain if I got even a whiff of the fumes."

Years after the onset of his illness, Ross learned that a dry-cleaning solvent had contaminated his town's drinking water supply. Ross investigated the possible connections between the contamination and his symptoms, and he heard about the Environmental Health Center in Dallas. "I didn't know if it was some kind of snake oil place, but I figured, 'What the hell.' I'd been very ill for years." Ross enrolled at the clinic as a patient. He was impressed by their explanation of his symptoms; some of their treatments made him feel better. He later returned with his family to Texas, signing on at the clinic as a staff physician.

Ross led us through the main corridor of the Environmental Health Center, pointing out the painstaking measures to protect his patients from inadvertent chemical exposures that might trigger illness. Before visiting the clinic, we had been instructed not to use deodorant or after-shave. We wore plain white cotton shirts and blue jeans washed in baking soda. Back in our hotel, we left the bright coral-green sweater that had distressed Dot Dimitri.

At the front of a special room, reserved for "the most severely sensitive patients," we watched a half dozen middle-aged men and women sit quietly in their metal folding chairs, waiting for their next battery of tests. When we took some photographs in the corridor, one of the patients in the special room suddenly fainted, dropping to the floor with a dull thud the moment the flash fired. A staff nurse calmly scooped her off the carpet and fit her back on the metal folding chair. Another employee urged us not to worry; it happened all the time.

At the far end of the Environmental Health Center, we reached the testing room. Ross explained that behind this laboratory's locked doors stood the apparatus that enabled his staff to scientifically prove the validity of chemical sensitivity. We entered the room, with great anticipation, to face what appeared to be nothing more than a pair of glass phone booths, without phones.

On the top of each rectangular glass booth, the staff had attached a vacuum suction device to clean the air and remove all possible contaminants. Each booth stood empty, except for a stainless steel stool. Much of the research underpinning the clinical ecologists' theories took place inside these simple glass enclosures.

Prior to entering the booth, patients studiously avoid contact with the chemicals they will encounter during the tests. This period of "desensitization," according to the clinical ecologists, allows the immune system to discard any adjustments it may have made to cope with the stress of everyday chemical exposure. After several days of desensitization, the patient enters the isolation booth and the round of testing begins with common substances such as formaldehyde, a chemical found in carpets and construction materials.

When Robert McIntyre entered the glass booth in 1977, he felt a vague sense of anxiety, knowing that he might soon confront some chemical that had caused him much distress in the past. The staff attached McIntyre to an EKG machine to monitor his heart rate. A doctor held up a small, sealed, glass laboratory vial, filled with a dilute concentration of an unidentified chemical. The test was double-blinded: Neither the doctor nor McIntrye knew whether the clear liquid was a synthetic chemical or a placebo. The doctor placed the vial on the floor of the booth near McIntyre's stool and sealed the door.

The vacuum cleaner atop the booth hummed. McIntyre reached down and removed the stopper from the glass vial. "I reacted instantly," he recalled. "I went crazy. I had an unbelievable headache, I got all uptight, my pulse and blood pressure shot way up." McIntyre flung open the door and ran out of the booth.

Ross told us that when he first entered the booth as a patient, his experience proved less dramatic, though equally persuasive.

"The thing that appealed to me," said Ross, admiring the empty glass chambers, "was that here I was in a specialized unit that used filtered air. It was exceptionally clean. It had an air lock. They were controlling as many of the extraneous and independent variables as they could. The isolation booths demonstrate clearly to the patients that the symptoms they've been feeling are brought on by chemical exposures."

One of the test vials Ross encountered in the glass booth contained the dry-cleaning solvent that had contaminated his drinking water in Nova Scotia. At first, Ross did not respond in any way to the fumes from this vial. "I thought it was a placebo," he said, "but an hour later, after I left the booth, I was badly affected. The joint pains, the muscle aches, I couldn't think straight. I had a headache and I was nauseated. It reproduced virtually all my symptoms. I felt so rotten, but I was pleased. There was finally scientific evidence of why I had become ill."

Scientific evidence of this kind fails to convince skeptics. Dr. Ephraim Kahn, a toxicologist at the California Department of Health Services, is one of the Environmental Health Center's leading critics, and he believes that the glass testing booths reflect the fundamental flaws of clinical ecology. Kahn bases his criticism principally on the discipline's overvaluing of subjective response. The glass booth tests involve no blood or skin reaction tests, nor any other objective measurement; the final proof of sensitivity is predicated on nothing more than what the patients say they feel. In truth, asserts Kahn, when patients are given blood tests for immunological abnormalities, or are observed in other ways to chart their reactivity to synthetic chemicals, nothing is found. Increased heart rate and blood pressure certainly indicate that the patient is alarmed, but it does not mean that the chemical has actually caused harm. The only uncontestable claim that clinical ecologists can make about their patients is that they feel bad.

Kahn asserts that the research of clinical ecology relies on the placebo effect—or the "suggestibility of human beings and the complex interaction of mind and body." Clinical ecologists, he argues, get their results in the same time-honored, but distinctly unscientific method of medicine men, shamans, and faith healers. "People with psychological symptoms such as depression or psychosomatic complaints," continues Kahn, "are especially vulnerable and may readily accept the diagnosis of environmental illness. What follows then for the more severe cases can be a true horror story. They live in constant fear of exposure to perfumes, detergents, antiseptics, plastics, sprays, printer's ink, and food additives." In desperation, notes Kahn, these people may even move to some isolated mountaintop, far from the modern world, where they will continue to struggle with their impossible lives.

- - -

At the top of High Lonesome, where the fresh breeze sweeps through the Davis Mountains and blows faithfully each day across the West Texas desert, Kari Pratt and Robert McIntrye felt their health slowly improve. And as they grew stronger, their relationship flourished.

"We're able to be together because of the good fortune that we react to the same chemicals," explained Robert. "I've never met anybody like that before. It's a blessing. We've been able to help each other through extremely difficult times."

"The thing I love about Robert," said Kari, "is his huge smile. He puts me so much at ease. It took me a long time before I trusted him. My whole life I'd been around abusive people. But Robert seemed so kind and different. He seemed genuine."

After years of living in the desert, Kari's pain had diminished and the anxiety that sometimes muddled her thoughts began to clear. "I really felt like this was the end of my illness." Each night, she climbed into Robert's one-ton 1979 Chevy pickup truck, spread her bedding out on the front seat, and drifted safely off to sleep. The threat of outgassing chemicals in her own trailer still concerned her. Robert's truck had covered over 350,000 miles, and the front seat was so old and worn that its synthetic seatcovers and the plastic knobs and dials adorning the dashboard posed no danger.

"Robert's got a real good car," affirmed Kari, admiring the rear bumper of the beat-up truck, where Robert had plastered a sticker declaring: *Life is Terrific!* "It's important that it's a Chevy and not a Ford. I've learned that." Over time, Robert's health also improved. He and Kari occasionally ventured down from High Lonesome to watch her son playing sports at the local high school. Before taking ill, Robert had coached several youth teams, and now he felt a renewed interest in basketball and football. He even went so far as to pick up an application for a coaching job. "Unfortunately," he recalled, "when I got to the school, they'd sprayed an insecticide in the gym. I never filled out an application."

Robert regarded the incident only as a minor setback. Most days, he adored his perch on High Lonesome. "The mountains here have all the good things," he boasted, gesturing across the vast expanse below us on a perfectly clear, blue sky afternoon. "They're the only place left in the United States that's got clean air. There's basically no agriculture here, there's no oil or gas production. People that can't live anywhere else can make it here. It's true that the most important thing for us is avoiding chemicals, but up here you can also avoid a

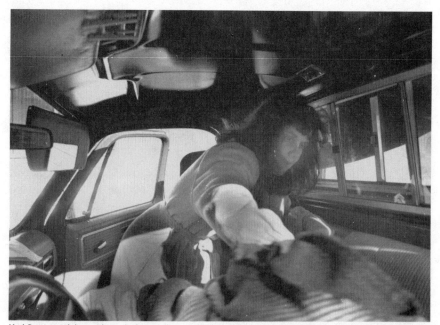

Kari Pratt, unable to sleep in her trailer, makes her bed in the front seat of Robert's Chevy truck.

lot of stress. I enjoy being here. I'm an optimistic and positive person, just plain glad to be alive."

On July 27, 1987, trouble came to paradise. Kari and Robert had just returned to High Lonesome after an expedition to town for supplies. "We went to Robert's house to unload groceries," remembered Kari. "Suddenly my skin was burning, my heart pounding, the abdominal pain—everything came back. I couldn't believe it. What in the world was going on around here?" Robert felt even worse. He thought he was having a heart attack. "We just got out of there," said Kari, pursing her mouth sternly in recollection of their terror.

Robert and Kari climbed into the front seat of the Chevy pickup, and drove 30 miles into the desert. Kari's children had gone back to Iowa to live with their father. So for three weeks, Robert and Kari camped out amid the sagebrush, trying to imagine what could have possibly gone wrong. Although they didn't know it at the time, a local rancher had started a new business 12 miles from High Lonesome, called the Blue Mountain Vineyard. Like most growers, he sprayed his grapes with pesticides. When Robert and Kari learned about the new vineyard, they felt immediately and absolutely convinced that the pesticides had risen through the air to their home high in the mountains, triggering their sudden illness.

Hundreds of miles away, at the office of the state agricultural department's Pesticide Regulatory Program in San Antonio, Ellen Widess and her staff shuffled through stacks of complaints from Texas farmers about the agency's overzealousness. Letters regarding unregulated pesticide dangers also piled up on their desks from farm laborers, chemical workers, consumers, ranchers concerned about the health of their cattle, suburban homeowners worried about their neighbors' backyard spraying. Widess had arrived in San Antonio one year earlier, after being recruited to head the department's Pesticide Regulatory Program by Texas' populist commissioner of agriculture, Jim Hightower. "Hightower had made a conscious decision to transform the pesticide division into a serious environmental, public health, and worker protection unit," she explained, "while also serving the needs of the Texas farmers. We were going to develop programs for pesticides that would be models for the entire country. I was an advocate for change."

When Widess read the letter postmarked Fort Davis, she groaned. Her backlog of cases numbered over six hundred. Her staff consisted of one lawyer, three investigators, and thirty-four field workers to enforce regulations covering 234 counties sprawled across the largest state in the continental United States. As an attorney who had formerly specialized in workplace chemical exposures, and whose clients

included several environmentally ill people, Widess immediately
grasped the legal, scientific, and emotional complexities she would
face atop High Lonesome.

"When I first heard from Kari and Robert," confessed Widess,
"it was not something I relished. I knew that once they latched on,
the investigation would be long, intense, and would probably never
resolve anything. Nevertheless, I tried to keep an open mind. I really
wanted to find out if pesticides in the Davis Mountains were the cause
of their problems." But Widess also knew the area's geography: The
Davis Mountains rank as the state's second-highest range, thrusting
off the flat desert floor to a height of more than eight thousand feet,
thus erecting a formidable barrier against extremely small amounts
of air-borne pesticides. "I couldn't think of any way that pesticides
from that vineyard could reach them way up there in the mountains."
But Widess found herself in a tough position politically. As the head
of a self-declared reform agency, she needed to pursue any serious
pesticide threat to public health. Yet the small vineyard at the foot of
the Davis Mountains proved exactly the kind of start-up family farm
that Hightower wanted to encourage in order to temper the political
clout and environmental effects of Texas agribusiness. Despite the
skepticism of her staff, Widess dispatched her best field workers to
the vineyard near Fort Davis. They found the grape growing area so
small that aerial and machine spraying had proved impractical; the
offending pesticides were applied by hand. The investigators obtained
records for all chemicals used in the vineyard. Then they headed up
the Davis Mountains, taking air samples in the territory around High
Lonesome. The tests failed to identify even the slightest trace of pesticide.

"Those people are nuts," complained one inspector, in his telephone report to Widess.

"No," she insisted, "I want you to test at different times of the
day, I want you to test at various distances from their homes, I want
you to be absolutely sure there isn't *any* pesticide up there."

The inspector spent three days in the mountains, carefully monitoring the air near High Lonesome. Again, the tests turned up no
pesticides. Widess felt convinced. When she reported the negative
results to Kari and Robert, they responded that their own bodies were
far more sensitive than the test equipment. If they reacted, there had
to be pesticides up there.

"I explained that the regulatory system, as we know it, does not
allow for that assumption," said Widess with a note of recollected
weariness. "Based only on their claims, I could not force the complete
economic disruption of a vineyard that was complying with all the

laws, using approved pesticides, and using them in a way that was careful and environmentally sensitive."

Kari and Robert concluded that Widess and her agency had joined in a conspiracy against them with the growers. Their demands escalated. Now they wanted the state legislature to declare the three counties surrounding High Lonesome to be the nation's first pesticide-free zone. They organized other environmentally ill people whose homes lie scattered throughout the mountains; together, they flooded the local media with complaints, passed around petitions, lobbied their representatives at the state house in Austin. Although Widess remained convinced that the Blue Mountain Vineyard was not responsible for the illness atop High Lonesome, she dutifully contacted the state health department and worked with its epidemiologists. Unsurprisingly, nobody could explain the illnesses—nor find any evidence of pesticide exposure.

"They got an enormous amount of attention from us," Widess admitted ruefully, "and there were so many other pressing issues in Texas that we had to deal with—farmworker safety, children's exposures. But Robert and Kari were so insistent. There was a momentum that they generated, and it just couldn't be stopped."

Dealing with the couple proved increasingly frustrating for Widess. No telephones existed on top of High Lonesome, and Robert and Kari feared exposure to paints and solvents in Fort Davis' public phone booths. "Yet they never stopped calling," Widess complained. "And since we couldn't answer them by phone, it just got worse." Then the vineyard owners started calling with complaints that Robert and Kari had turned them into villains, when they had done nothing wrong.

Although pesticide use continued at the Blue Mountain Vineyard, Robert and Kari moved back home to High Lonesome. They corresponded regularly with the local newspaper, threatening lawsuits and raising as much controversy as seemed humanly possible from one of the most isolated peaks in the state. The vineyard's original owners gave up and sold their operation. Then Kari tried a new tactic. She wrote a personal letter to the new owners, explaining in the most moderated, rational tones why she believed their vineyard pesticides endangered her life.

"I had somebody edit the letter over and over to get all of the anger out," she said. "I just let them know how sick we were."

Soon after Kari sent her letter, the Blue Mountain Vineyard went organic—for reasons that the new owners never made clear—curtailing all pesticide use. "It's wonderful," declared Kari, beaming with enthusiasm. "We're not getting exposed, and they're showing you can grow grapes here without using toxic chemicals."

Although Ellen Widess firmly believed that pesticide use should be reduced throughout the state, the final outcome in the Davis Mountains bothered her. "Having a vineyard go organic is a positive result," she admitted. "But the steps it took, and the people who forced it—the lack of credibility leading to the solution. . . ." Widess sighed and threw up her hands. "It's just not a good model to apply to other places." Today many staunch environmentalists privately concede that small victories engineered by the environmentally ill from atop places like High Lonesome do not advance the cause against chemical contamination waged by citizens in Yellow Creek, Kentucky, or Yukon, Pennsylvania, or Spencerville, Ohio, or McFarland, California, or countless other communities. "Even for those of us who are sympathetic with the environmentally ill," insisted Widess, "there's nothing that can be done that would ameliorate their condition that's practical, that's feasible, that's economical, that's fair."

Robert and Kari, she believed, would always remain isolated from the larger struggle, stranded on top of High Lonesome.

- - -

Nobody can state for certain whether environmental illness truly exists. But for the first time since the phrase was uttered 40 years ago, the controversy is finally progressing beyond the squabbling of various true believers.

In 1990, the National Research Council, the investigative arm of the National Academy of Sciences, acknowledged the importance of placing the disease under scientific scrutiny. In essence, the National Research Council assumed responsibility for mediating between the skeptical medical establishment, composed largely of allergists and psychiatrists, and the insurgent ranks of clinical ecologists and their patients. Researchers and clinicians from both sides of the controversy convened for a national meeting, and the concept of environmental illness immediately took one giant step towards respectability.

Mark Mendell, a researcher with the California Department of Health Services who has reviewed much of the literature regarding multiple chemical sensitivity, attended the national meeting. He concluded that the allergists and other skeptics showed up only under strenuous protest, believing that the convocation was a ploy to align the prestigious National Research Council with a nonexistent malady. "These anti-multiple chemical sensitivity people feel so strongly that there is no possibility the illness exists," emphasized Mendell, "they are so angry at the clinical ecologists, that they cannot separate what the clinical ecologists say and do from what might be occurring with their patients."

Mendell attended one caucus charged with devising a clear definition for the illness. "This working group had all the enemies in it," he recalled, "all the clinical ecologists and allergists who wanted each other dead. They spent a day and a half in the same room, yelling, not listening, and ignoring each other." The group finally agreed on one thing: They needed a study that compared patients with multiple chemical sensitivity to patients with chronic illnesses, such as emphysema or arthritis. In this way, the psychological state of the environmentally ill could be contrasted with unarguably sick people, indicating whether the odd behavior of many chemically sensitive patients resulted from their chronic and untreatable illness or their deep-set emotional problems.

But an insurmountable problem blocked their progress. At least one hundred people with multiple chemical sensitivity were needed for the experimental group. And while their chronically ill counterparts in the control group could be easily identified, no test existed to spot environmental illness. Once again, the truth regarding environmental illness seemed to be chasing its own tail. Scientists had yet to agree, even in the most general terms, on the malady's definition; therefore, nobody could separate the truly ill from the crazies, malingerers, or merely misidentified. An experimental group of chemically sensitive patients might mix the medical histories of mind-reading Dot Dimitri along with the far more rational Dr. Ross, and then fill out the ranks with Robert McIntyre, Kari Pratt, and numerous other exclusively self-diagnosed sufferers. Even the clinical ecologists admitted that *some* "chemically sensitive" patients suffered largely from emotional distress. But nobody could tell exactly who these people were. As a result, serious researchers could not even pose the central questions regarding human physical response to minute quantities of everyday chemicals.

Recently the National Academy of Sciences offered to fund research in immunology, psychology, toxicology, and other areas that might eventually advance the understanding of environmental illness. And even the Environmental Protection Agency has agreed that definitive answers to the puzzling questions regarding multiple chemical sensitivity must be forthcoming. In a letter to the National Research Council, the director of the EPA's Indoor Air Division wrote that his own agency ". . . will have to make an assessment . . . as to whether such a syndrome exists and, if so, how affected individuals should be

In the clean air of Wimberly, Texas, Sue Pitman built a house in which she could live virtually outdoors.

accommodated." Given the EPA's doleful record in charting the effects of many thousands of known toxic substances, everybody involved in the controversy must be prepared for a long wait.

-    -   -

Before driving up to High Lonesome for our final meeting with Robert and Kari, we stopped at a restaurant in Fort Davis to purchase a picnic lunch. Unfortunately, the burgers-and-milkshakes menu failed to accommodate the strict diet of the chemically sensitive. We reluctantly settled on a salad bar mix of iceberg lettuce, carrot sticks, and golf ball radishes. Unable to locate a paper bag, the woman at the cash register stuffed the salads into clamshell Styrofoam boxes. Robert and Kari waited for us at the top of High Lonesome. Before we could apologize for the probable pesticide residue of the nonorganic salads and their petrochemical, eternal-life containers, our hosts gathered together four chairs on the porch, flipped open the clamshells, and launched into a binge of ravenous munching. Together we gazed out across the vast unraveling plain, eating in silence as the late afternoon sun cast a precise line of shadows from the ragged mountain slopes down to the desert floor. We didn't have to ask why the food and its containers did not offend, alarm, or even worse, harm our hosts. By the time we reached Texas, we had already visited contaminated communities in a dozen other states. We had spoken with scores of people who had proven deeply, often frantically, concerned about the chemical threat to their homes and families. The environmentally ill offered the most extreme example of the inconsistencies and confusion that characterized life in hundreds of toxic towns scattered throughout the country. In the face of unpredictable risk, cool reason and consistency often proved impossible to muster.

Finally, we asked Robert and Kari one last question: What would they do if their illness was suddenly cured?

"I would leave here," Robert spoke up immediately. "I would like to have a normal life, but I've learned the hard way that you have to be very careful. I'd avoid any significant chemical exposures that might bring this all back. But I would leave this very place where we're sitting today"—he gestured across the unblemished horizon—"and get back to my work in whatever way I could."

Kari didn't answer immediately. Finally she said softly, "I've always dreamed of living in the mountains. And having somebody like Robert who smiles when he sees me. Even if I was cured," she insisted, gazing out across the desert of falling shadows, "I'd stay right here."

Months later, after returning home to California, we received a greeting card from Kari and Robert. The cover featured a grinning Snoopy, roasting marshmallows over a camp fire. The balloon above Snoopy's head read: "Thanks. You made me a happy camper."

Inside, Kari had sketched an elegant drawing of her mountain home. But hovering over the roof of her trailer, the drawing also revealed an ominous, darkly shaded, mushroom-shaped cloud. In the middle of the cloud, Kari had etched in bold capital letters: **ON AUGUST 11, 1991, A HUGE INVISIBLE CHEMICAL BLANKET MONSTER FORCED US TO LEAVE OUR HOMES AGAIN.** In tiny, almost unreadable letters, she had squeezed the following words into the blank space at the bottom of the card: "lungs burning, head burning, eyes burning, very sick-awful feeling all through my body, extremely severe abdominal pain, pink itchy rash on thighs. Deathly ill."

On the opposite side of the card, Kari had penned a more personal note.

"Robert and I thoroughly enjoyed your visit. We appreciate very much your interest in our dilemma. And thanks again for the great lunches.

Kari Pratt.

"We both became deathly ill after the campground west of us sprayed fenthion. We had to leave our homes for six weeks. We are fighting hard to again recover the ground lost on our health.

"Please, keep in touch.

Kari"

"Around Here, You're Either on One Side or the Other!"
—McFarland, Summer 1988

Maria Puentes, a mother from McFarland, picks grapes on Pandol Brothers' farm.

By the winter of 1988, McFarland was more fiercely divided than anybody had previously thought possible.

The opposing sides were represented by Connie Rosales, the woman who first sounded the alarm about the cancer cluster, and the United Farm Workers, the organization most deeply involved in combating pesticide dangers in the fields. On the surface, their enmity appeared odd, almost perverse. After all, both sides had agitated vigorously for a solution to McFarland's cancer cluster mystery. But irreconcilable differences in tactics, loyalties, politics, personal styles, and expectations made their logical alliance untenable.

The central event precipitating the split between Rosales and the UFW involved the union's 14-minute video, "The Wrath of Grapes." The video, in which Rosales and several other McFarland families briefly appeared, now served as the centerpiece for the union's publicity and fund-raising strategy to promote a new grape boycott. By 1988, the UFW had distributed fifty thousand copies of "The Wrath of Grapes" throughout North America. One year later, another twenty thousand copies found their way to more than 40 U.S. and Canadian cities where UFW supporters coordinated the campaign against pesticides. From the start, Rosales had presumed that the video would dwell chiefly on McFarland's cancer cluster; in fact, it dealt more generally with the abuse of pesticides throughout the Central Valley—while building the case for the union's latest organizing tactic, the grape boycott.

"When I found out that "The Wrath of Grapes" supported the grape boycott," declared Rosales, "I flipped out."

For months, Rosales railed against the UFW and its ally, the National Farm Workers Ministry, standing up at public meetings and in media interviews to proclaim that both groups had betrayed promises to assist McFarland's cancer-stricken families with money to defray medical, living, and burial expenses. The UFW and National Farm Workers Ministry officials denied the pledge, though they had indeed helped several families out of compassion, rather than obligation. But Rosales' chief complaint seemed to be that the union had exploited the personal tragedies of McFarland's children in order to pursue its own political agenda.

"When my son came down with cancer," said Rosales, "I was in shock. Then the UFW came in and said, 'The growers are killing you with pesticides.' And I'm going, 'Yeah, yeah, you're right, that's true!' I'll never forgive the UFW for that. For using me when I was so vulnerable."

Yet Rosales' denunciation of the UFW could not simply be dismissed as a quixotic personal vendetta. The discord in McFarland highlighted a long-standing, if muted debate over the UFW's early promise and recent performance that now divided many small agricultural towns in the region.

Over a period of 16 years, UFW organizing had boosted farm workers' real wages by 70 percent. For the first time in American labor history, field laborers earned pensions, disability insurance, health care benefits, and access to a credit union. These accomplishments constituted the union's glory days, now long past. Since the late 1970s, the UFW's fortunes had plunged, as numerous contracts unraveled and widespread support among workers dissipated. But whatever the union's difficulties, it still remained a powerful source of controversy within the Central Valley.

"Around here, you're either on one side or the other," declared Sister Pat Drydyk of the National Farm Workers Ministry. "You can't be in the middle. There's no way, no way."

Nevertheless, as the split between Rosales and the UFW grew more acrimonious—and public—most McFarland residents strained to remove themselves from the debate. Despite their appreciation of the critical gains won by the UFW for farm laborers in the past, many people still resented the union's intrusion into the personal tragedies of the cancer-stricken families. Besides, moving to McFarland and buying a solid, stucco, three-bedroom tract home had represented an important step up for many Hispanic families who had once labored in the fields. Few McFarland residents wished to return to the ferocious battles between workers and growers that had characterized the early days of farm labor organizing in the 1960s.

"We need dialogue with the growers," insisted Rosales. "These guys are not total barbarians. They don't want to kill our children, but they also don't want to give up the farm. Agribusiness is the game here, and we have to help them make the transition to not using so

Connie Rosales.

many of the chemicals. We don't want to be just a tool for union propaganda.''

Yet some UFW supporters complained that Rosales had begun to sound like a propaganda tool herself—only for the growers.

In 1987, Rosales flew to the national conference of the United Methodist Church in Washington, D.C., her flight paid for by a local anti-UFW group called the Grape Workers and Farmers Coalition. At the conference, she opposed the union's pitch for support of the grape boycott. The Methodists endorsed the action anyway, but Rosales expanded her reputation as an implacable UFW foe. When she appeared in the pages of *California Farmer*, posing in a color photograph next to some grape vines, the split between McFarland's most prominent advocate for the children with cancer and the UFW appeared irredeemable. "They really don't care about the children who have cancer," Rosales declared of the union and ministry. "They just want to get sympathy and move people to send in donations for the UFW's grape boycott. . . . But in all the time they've been active in McFarland, I haven't seen them even buy a kid a teddy bear."

Relations reached their low point when Rosales carted into a McFarland community meeting a batch of refreshments that included table grapes—the object of the UFW boycott. Rosales mocked the notion that the grapes represented another jab at the union.

"God," she scoffed, "I was in charge of refreshments. Everybody around here eats grapes. I'd baked cookies, and took over coffee and tea—and I brought grapes and kiwis, too. But everyone around here acts like I committed some kind of crime." Indeed, for several years thereafter, UFW supporters bitterly recalled the gesture's symbolism.

Yet some outsiders observing this rivalry refused to read into Rosales' actions any precise political intentions. Nor did they believe Rosales had definitively aligned herself with the growers, or even less—as the rumors ran—joined their payroll.

"Connie was not political," insisted Dr. Lynn Goldman, who headed the McFarland studies for the California Department of Health Services. "She was a very naive person."

In fairness, even the most adroit political operatives had found the swift currents of the Central Valley's political waters difficult to maneuver. And in recent years, the scene had become even more complex, due largely to the rise and fall—and rise again—of the UFW.

The 1980s had worn hard on the union.

Back in 1976, the UFW had won over two hundred elections among farm workers who wanted the union to act as their bargaining agent with the growers. Six years later, the number of election victories dropped to eight. By 1985, the union had only three contracts with grape growers—down from fifty at its height. As the organizing of new workers slowed, the standing membership also dwindled. According to the *Wall Street Journal*, only 40,000 of California's estimated 350,000 agricultural workers belonged to UFW by 1986—down from a peak of over 100,000 in the mid-1970s. Union officials and growers disputed the precise numbers, but whatever way they cut them, membership was shrinking, not expanding, as had been predicted in the 1960s, when North America's entire agricultural sector—from British Columbia to Florida, from Alberta to Texas—was expected to eventually organize along UFW lines.

Cesar Chavez (the union's founder and president) and other UFW officials attributed full responsibility for the union's woes to the aggressive manipulations of California's Republican governor, George Deukmejian—the successor to longtime UFW-supporter, Governor Jerry Brown. Strongly backed by agriculture, Deukmejian had zealously repealed union advances won throughout the 1970s, frustrating efforts to both organize in the fields and administer current contracts.

But the Deukmejian administration's hostility took a greater toll because of convulsions and power struggles within the union itself. While the *New Republic* was still characterizing Cesar Chavez as "the last representative of a dying breed—the charismatic 1960s hero," people closer to the union described a far less flattering portrait.

Over time, Chavez had developed a reputation as something of a mysterious recluse, holing up at the union headquarters in the desert mountains. The UFW called its new home La Paz, "the peaceful place." (In an unexpected literary allusion, growers snidely deemed the former tuberculosis hospital, Chavez's "Magic Mountain.") And in a way, the headquarters was *too* peaceful: One-third of the central staff had been fired in 1977. Many defections followed, washing away the cream of the veteran organizers and legal staff. Chavez maintained that everybody could be replaced, but this concentration of talent never really reappeared.

In fact, the UFW seemed to be shifting strategies. And it drew much fire from ex-staffers and longtime allies for spending less time, energy, and money organizing workers in the fields than in cultivating support for the new grape boycott through a vast program of national fund-raising directed at the urban elite. Initiated in 1984, the boycott resurrected the UFW's single most successful tactic, which during the late 1960s saw 12 percent of the population—17 million Americans—boycotting table grapes. During these years in New York City alone, sales plunged 90 percent. Prices dipped by a third.

But the second time around, the UFW crusade against table grapes failed to catch fire. A 1986 *Wall Street Journal*/NBC News poll found that 77 percent of the people questioned didn't even know about the boycott.

What the UFW required was a new twist on an old angle, which turned out to be a long-standing, but little emphasized feature of the union's platform: the dangers of pesticides—and in particular, their potential threat to consumers of fruits and vegetables.

"For quite some time, the UFW had been looking for a way to build up its base again," said Rueben Garza, McFarland's first Hispanic mayor, whose farm-worker family of ten had migrated to the Central Valley from Texas in 1960. "The union had lost its luster and its strength in numbers. I suppose at that particular time, they felt that the environment would help their cause."

Indeed, the union's new tactic, prominently featuring McFarland's ailing children, breathed life into the boycott.

"What do you and a farm worker have in common?" asked one fund-raising letter. "Exposure to poisonous chemicals! . . . And there is conclusive evidence that you don't have to be a farm worker to be affected . . . *seriously* affected . . . by chemicals used in agriculture. *All you have to do is eat!*" In the spirit of the self-absorbed 1980s, the campaign stepped beyond the fundamental appeal to justice that had motivated the 1960s boycotters, instead invoking a rational argument of enlightened self-interest. Laboring over a Sperry 1100/70 mainframe computer and high-speed printing equipment, the union spewed out a million pieces of mail per month.

In some fund-raising letters, the UFW claimed that "Children in towns like McFarland are showing childhood cancers at 8 times the normal level." Later, the letters would up the ante to 12 times the normal rate. The California Department of Health Services still held the count at 13 cases, with 5 deaths—about 4 times the expected rate of illness.

As plaintive messages regarding McFarland's cancer cluster fueled national support for the grape boycott, several families appearing in "The Wrath of Grapes" moved to the courts in an attempt to block the video's further distribution. In June 1989, a California judge granted an injunction against disseminating and showing the documentary, despite a detailed affidavit filed by its producer swearing that all participants had been informed of the UFW's intention to promote the boycott by tying it to the issue of pesticide dangers. Around this time, one parent, more out of pique than conviction, went as far as to privately suggest that the cancer cluster's real culprit might have been the UFW's $5 million microwave communication system linking field offices with the union's headquarters in La Paz.

But the cancer-stricken families were not the only people feuding with the UFW. One of the most surprising ruptures occurred between the union's scientific advisors and Dr. Beverly Paigen, the epidemiologist who for a time sat on the governor's Scientific Advisory Panel on McFarland, and whose ground-breaking studies of Love Canal had led her to work alongside community groups trying to prove that they, too, had been made ill by toxic exposures.

"I always liked the UFW," said Paigen, "but I think they really are using McFarland. We once had a conversation about which pesticides I thought were responsible for the cancers, and the UFW was upset with me because they weren't pesticides used on grapes. Well, great—but I can only say what I think it is. Their interest is less in McFarland than it is in making their point about pesticides."

Paigen had suspected dinoseb, a pesticide linked to birth defects. Marion Moses, of the National Farm Workers Health Group and president of the Pesticide Education Center, agreed that dinoseb was an extremely dangerous compound. "It is much too toxic," stated Moses. "Employers do not protect or instruct their workers, and doctors do not know how to treat poisoned workers. Because it is slowly excreted from the body, the margin of safety is much too narrow as workers can accumulate potentially lethal amounts of it in their bodies. Because it also contaminates the groundwater, we have to be concerned about the effects on all rural residents, not only farm workers."

But dinoseb had been banned in 1986. And if dinoseb turned out to be the culprit, as Paigen believed, the discovery would be of little political use to the UFW. If anything, the state would probably take credit for its successful vigilance against the pesticide—despite the damage done. Paigen believed that the union opposed her ideas about dinoseb on political, not scientific grounds: In order to bolster its national crusade against pesticide dangers, the UFW required a chemical still used in the fields to be the guilty substance.

"The UFW has taken a very active role in the cancer cluster," admitted Paigen, "but many of the local people feel the union is ripping them off—in the sense that it's their suffering, and the farm workers are getting the political benefits."

Without doubt, McFarland was having enormous difficulty gathering together the kind of community group that had emerged successfully in other towns to pressure local, state, and federal officials. At one point, Penny Newman, now working as a regional organizer for Citizen's Clearinghouse for Hazardous Wastes, responded to a request from McFarland residents to try to help the community organize itself, but with no lasting results.

Yet some inconstant efforts did make waves. On June 16, 1988, three hundred people attended the first meeting of United McFarland Community, a new grassroots organization concerned with "the cancer problem, its cause, prevention, and cure." Sister Pat Drydyk, of the National Farm Workers Ministry, helped set up the organization, which she characterized as "people organizing themselves and dealing with issues of pesticides, water, voter registration, and housing." The UFW also played a key role in forming the group. Cesar Chavez addressed the first meeting.

"Only the union has been willing to talk about *banning* some of the pesticides," emphasized Sister Drydyk. "We all have to join in. We don't want to wait until they've counted the bodies."

United McFarland Community met several times, demonstrating some initial promise. At one point, its leaders announced that Latino inmates at the state prison in Vacaville had donated $700 to the organization from proceeds raised through a bake sale. Despite this unlikely endorsement—and more importantly, the heavy turnout for the first meeting—United McFarland Community never became a driving force within town. Nor did the majority of McFarland citizens most concerned about the cancer cluster ever identify the group as their advocate or representative. Actually, the greatest amount of political heat, if not always light, was being generated by people outside of town, who for various reasons temporarily embraced McFarland's troubles as their own.

On May 25, 1988, the people of McFarland were astonished to find the national press converge upon them as Jesse Jackson's campaign for the Presidency of the United States stormed through town with its resounding message of "Keep Hope Alive!"

In McFarland, Jackson caucused with the parents of the cancer kids, spending the night at Tina Bravo's home. "It was so exciting," exclaimed Bravo's ten-year-old daughter, Yadira. "He had a Secret Service man standing at his door. Almost my whole class came to see him last night."

Not everybody in town welcomed the national spotlight. The week before, during an impromtu speech, Jackson had mused at the microphones about McFarland's "contamination and corruption." Local officials feared similar condemnations might erupt live on network news. But once Jackson appeared in town, accompanied by McFarland's mayor, vice-mayor, water company manager, and the parents of the cancer children, he retooled his criticism to take on the state and federal authorities.

"The idea that there is contamination and corruption is self-evident," declared Jackson, as the crowd gathered around him at McFarland's well number five, which had been closed following tests revealing high levels of the pesticide DBCP. "It does not reflect on the local officials, but certainly the people here have been abandoned." The politicians and business leaders flanking the candidate appeared visibly relieved.

Alongside Jackson stood another influential McFarland resident, Martha Salinas. Over the past several months, Salinas had emerged as a prominent figure in public meetings—sometimes eclipsing even Connie Rosales, as she heatedly engaged the state health officials in endless debate. Quick-witted, telegenic, and verbally deft, Salinas now provided the sound bites to television news and newspaper reporters that had been Rosales' province before her painful dispute with the UFW dampened some of her enthusiasm for the fight. Standing in Jackson's shadow, Salinas was now reaching a national audience for the first time. As the cameras rolled at well number five, Salinas refused to let the local leaders off as easily as her mentor had.

"You say you are so concerned about McFarland!" she admonished the city officials, her voice ringing out in disbelief. "What about bottled water for our school children?"

During the rally, Jackson compared Martha Salinas to Rosa Parks, the African-American woman from Alabama, who in 1955 refused to give up her bus seat to a white rider—thus sparking the Montgomery bus boycott, a watershed event of the civil rights movement. In reality, the similarities between the two women seemed remote. Salinas had grown up in the Central Valley during the 1950s, picking cotton and watermelons as a young woman. "I can't work in the fields around here anymore," she said. "A lot of the growers don't like me because of the stand I've taken. They won't give me work." But some McFarland residents, particularly the Anglo skeptics who stubbornly maintained that the town's health problems were highly exaggerated, regarded Salinas warily.

"It was very difficult to challenge her," complained science teacher Merry Ellen Alls. "If you did, you were against mom and apple pie. The men had a particularly difficult time because they were challenging a woman defending her children."

In fact, Alls lamented that Salinas often showed up at public meetings with her daughters in tow, speaking in a manner that prompted listeners to infer that her kids suffered from some mysterious, if unverified illness. "At one meeting," recalled Alls, "she stood up, saying 'our kids are dying.' And her kids were standing right there beside her."

McFarland resident, Martha Salinas.

One of the ideas Salinas had most recently championed was opening a health clinic for McFarland's low-income families—a notion also espoused by Jackson in his address at well number five.

"Every time a child gets a headache or a joint ache, everybody immediately thinks of cancer," Jackson told about 60 reporters at the well. "A clinic is necessary for peace of mind, but also for early detection of cancer."

Jackson's confusion over the efficacy of screening children for cancer would soon spread throughout the community, further muddling the scientific and public policy questions framed by the cancer cluster mystery. But for the moment, his words inspired many McFarland parents.

When the national media departed McFarland, the town lapsed into an uneasy quiet for about two weeks. Then, on June 4, 1988, Jackson's presidential campaign swung back through McFarland for a second appearance—this time, in full regalia. On a bright spring day, the Rainbow Coalition's charismatic candidate cut a bustling swath through the center of town, his arms linked with UFW president Cesar Chavez on one side and McFarland mother Martha Salinas on the other, marching determinedly to address a rally of 2,500 people.

"Why are we marching today?" Jackson demanded of the crowd. "To save the children. To stop the cancer. To make our food secure."

Jackson wore a white San Francisco 49er's polo shirt and a button reading, "Boycott Grapes." Behind the candidate, Chavez, and Salinas trailed an entourage of celebrities recently arrived from Los Angeles and the San Francisco Bay Area. Flanking Jackson at the rally were actors Ned Beatty, Susan Sarandon, Margot Kidder (who played Lois Lane in the Superman films), Carl Weathers (who appeared as the boxer Apollo Creed in the *Rocky* series, and whose current movie title, *Action Jackson*, became the candidate's campaign moniker), Kim Fields (of television's "Facts of Life") and singers Bonnie Raitt and Holly Near. The Central Valley had not witnessed such a convergence of celebrities since the heydays of the late 1960s, when Cesar Chavez marched through Delano's fields alongside Robert Kennedy in support of the first grape boycott.

But Jackson's entourage did not consist entirely of recording and screen stars. At the podium, next to Jackson, stood Luz Gallegos of Delano, whose four-year-old daughter, Rosalene Lagutan Gallegos, had died one week earlier of adrenal gland cancer.

"Stand tall, McFarland, and we'll stand with you!" shouted out Jackson, as Gallegos stood silently at his side.

The candidate's voice croaked hoarsely, worn from the heavy campaign schedule. Nevertheless, he waved his hands in the air and pounded the makeshift podium, energetically informing the crowd that their community deserved "an Environmental Protection Agency that will protect the environment, not the agency." The people of McFarland cheered. Jackson called for a federally financed investigation of the cancers' cause, a free clinic for treatment, and a ban on untested pesticides. Applause rang out among the spectators.

"We are here because children are dying," repeated Jackson. "The government says, 'Prove it.' Let them go to the funerals. Don't tell parents to prove anything. Children are dying, that's why we're here. We need money and health services in McFarland, and we need a president in Washington—and I'm looking for a job!"

By the time Jackson, his campaign staff, the press, and the celebrities blew out of town for the last time, McFarland was fixed firmly as a site worthy of political pilgrimage.

Lieutenant Governor Leo McCarthy, a longtime aspirant to the governor's seat, lobbied the Deukmejian administration for money to clean up all contamination uncovered in the town, supply bottled water to school kids, and sponsor a health clinic. In October, two weeks prior to the presidential election, Jack Parnell, director of the California Department of Food and Agriculture, met with Jack Pandol, Sr., the county chairman of the George Bush presidential election campaign, in one of Pandol's vineyards. Posing for the cameras, he gulped down an unwashed grape. "We do extensive testing and have for a couple of years," Parnell insisted. "Table grapes are absolutely the safest thing you could possibly eat."

Throughout this busy political season, Cesar Chavez also captured considerable attention. On July 16, 1988, the UFW president commenced his 36-day "fast for life" to protest the continued use of dangerous pesticides. With Chavez fasting, like in the old days, numerous public figures rejoined the cause. Members of the San Francisco Board of Supervisors endorsed the grape boycott. Atlanta mayor Andrew Young proclaimed his own intention to fast. Twenty-three days into the fast, Chavez appeared before television cameras to reaffirm his commitment to stick out the long ordeal. Accompanying him were actors Robert Blake of television's "Baretta," Charles Haid, who portrayed Renko on "Hill Street Blues," and Luis Valdez, director of the hit movie *La Bamba*, whose stage career had commenced with Teatro Campesino, the agitprop theater company formerly affiliated with the UFW.

When Chavez finally broke his 36-day fast on July 16, 1988, the public ceremony once again invoked the spirit and symbols of two decades past. Following a Mass held in a large tent outside of Delano, Chavez accepted a piece of bread from the hand of Ethel Kennedy—just as he had done in 1968 from her husband, Robert Kennedy, in curtailing a 25-day fast. Twenty years earlier, Robert Kennedy, then a candidate for president of the United States, had pinned a UFW button to his lapel and declared his support of *la huelga* and *la causa*—the strike and the cause of justice for farm workers. Now Kennedy's widow and the couple's three grown children, Kerry, Rory, and Christopher, issued their own emotional statements.

"The day will come," proclaimed Kerry Kennedy, her voice strong and sharp, "when grapes are no longer washed with the tears and the sweat and the blood of the workers of these fields."

Too weak to make a speech—Chavez had lost more than 20 pounds—the UFW president asked his oldest son, Fernando, to read a prepared statement, thanking supporters and reaffirming his commitment to the boycott. Chavez referred to his fast as "a fervent prayer that, together, we will confront and resist, with all our strength, the scourge of poisons that threaten our people and our land and our food." One of Chavez's grandchildren presented him with 36 balloons, one for each day of the fast.

Back in McFarland, Connie Rosales remained unmoved. "The emphasis is being taken off our children and being put on him," she declared. "I don't have much respect for him."

Around town, other adversaries of the union recited a litany of uncheckable, and no doubt apocryphal, stories: Chavez was sipping milkshakes after sundown, as he had 20 years earlier during his first fast, and had purchased trousers two sizes too large to give the appearance of wasting away. The cynicism directed towards Chavez seemed cruelly personal and eccentrically emphatic. Some people in town appeared to be viewing every action related to the cancer cluster with growing suspicion.

Even Ron Huebert, the school superintendent who had skillfully charted a middle course between the outraged mothers and the skeptical officials, seemed terribly weary of McFarland's expanding notoriety. Huebert complained that part of the town's problems originated with the steady stream of outsiders, the organizers and politicians who announced their arrival with a press release. He characterized the outsiders as "human tornadoes," who blew into town, stirred everything up, and then disappeared in a dusty cloud. After Huebert organized a community prayer meeting, christened "Hands Around McFarland," a reporter asked if the school superintendent had finally shifted his hopes for solving the town's mystery from science to faith.

"Our Creator creates the scientists," Huebert answered. "So I would say a little mutual work there."

All the upheaval in town had led science teacher Merry Ellen Alls to severely doubt the judgment of her neighbors.

"I don't understand why these people don't trust anybody," she complained. "I trust the government. But they trust Connie. Why? Or some farm worker without any education will stand up and say, 'I know it's pesticides that gave my kids cancer,' and everybody will believe him. You could have a scientist, who has done all this work, and he'd say, 'It doesn't work this way'—and nobody would believe him. It's very difficult to tell what people understand here."

-　　-　　-

In December 1988, the McFarland Community Health Center opened its doors.

The California state legislature had originally committed itself to funding the low-income clinic, but Governor Deukmejian vetoed the $333,000 appropriation. Left on its own, McFarland's civic and business leaders raised $90,000 in private donations and pledges to begin services one day per week. Credit for turning McFarland's daydreams about the clinic into reality largely belonged to school superintendent Ron Huebert. To kick off fund-raising, Huebert pried a $36,000 donation from Contel Telephone Service, plus a $5,000 gift from the Pandol family, the area's largest grape growers.

"We've identified a lot of people throughout the state who really care about McFarland," said Huebert. "There was a time when this town thought nobody cared."

On the day of the clinic's opening, media and state political figures, including Lieutenant Governor Leo McCarthy, swarmed around the doorway, scaring off all but six families who had arrived for medical treatment. Over time, the clinic steadily picked up business, eventually serving 25 to 30 patients every day.

"I painted the trim myself," declared Steve Schilling, director of Clinica Sierra Vista, the health consortium that ran the clinic. "My brother-in-law and I poured the cement for the ramp out back. We took this little junker house, and turned it into a clinic in 30 days. We had no corporate plan, no scheme."

Most residents agreed that the clinic was one of the best things to happen in McFarland for years—and without doubt, the most positive outcome of the battles still raging over the cancer cluster. Forty-four percent of McFarland residents had no health insurance, and the clinic seemed perfectly suited to address their needs.

"When you take a population that is unserved and scared and powerless," insisted Schilling, "and then you bring them a little piece of service, they're just thrilled to death. They've got something to take their kids to."

Less than two months after the new clinic opened its doors, the state declared its intention to provide McFarland with another unexpected health service. Residents greeted this second announcement more equivocally.

On January 11, 1989, Dr. Ken Kizer, director of the Department of Health Services, announced that the state would "be sponsoring a child health screening program" from February through April. Prior to informing the community, Kizer had made an official announcement to the press. The first the community heard about the health screening program was when reporters called parents to ask what they thought about it.

The entire affair harkened back to the earliest days of terrible communication between the state and the community. And once again, it aroused suspicions in McFarland that the officials' first goal was good publicity rather than good public health. In an editorial, *The Record* in Delano railed bitterly: "It was our belief that we had progressed beyond this point in handling major announcements in McFarland. What about promises . . . that the parents affected and residents would be informed first? Has this policy been abandoned or was this merely a slip of the word processor? If this seems like deja vu, you're right. We're right back where we started from in regards to public information. While few can fault the department's investigative effort, it is certainly clear that health services has much to learn about public relations."

The department's haphazard announcement procedure also fueled confusion among McFarland residents about the precise meaning of "health screening." What the state officials meant was their plan to house medical staff in some trailers in the middle of town and then conduct a free, one-time routine physical checkup for every child from kindergarten to sixth grade. The health screening was the equivalent of a single visit to a doctor's office.

Yet many people in town misinterpreted the routine checkup as "cancer screening."

"Everybody thinks it's a clinic that's testing the kids for cancer," said Martha Salinas, three months into the program. Ron Huebert also noted that the "purpose of the screening is to assess the general health of the children, with an eye towards cancer early warning and early detection." *The Record* joined in the confusion, reporting that the department would "screen McFarland children for cancer" in a program that might "even lead to some clue as to what has caused the mysterious cluster." Even George Johnston, the administrative director of McFarland's new clinic, agreed that the program was "a one-shot cancer screening. One time and then they'll be gone."

Unfortunately, no way existed to identify "early warning signals" for the cancers that routinely struck children.

Cancers can be "screened," or detected in their early growth stages, only if two conditions prevail. First, the cancers must be of the type that prove relatively slow growing and tend not to spread rapidly to other tissues. Second, there must exist safe, practical, and relatively inexpensive methods for testing the patient.

These requirements often can be met for adults. PAP smears can indicate cervical cancer, mammograms may reveal breast cancer, and tests for blood in the stools might point to cancers of the colon. Cancers of the testes and skin also are amenable to early diagnoses.

But children don't get breast, colon, cervix, testicular, and skin cancers. They are much more often afflicted by cancers of the kidneys, blood, adrenal glands, brain, and bones. And since these cancers spread rapidly, a detectable "early growth" stage is practically nonexistent. Only by conducting examinations every few weeks could a doctor be relatively certain that a child remains cancer-free.

McFarland's confusion about the possibilities of early detection for childhood cancer disturbed Dr. Richard Kreutzer, who headed the state's health screening program. "The parents may be lulled into this false sense of security that a one-time certification of health means that the child can never get sick," said Kreutzer. "That is the inherent danger."

Yet the parents' misconception of the screening's purpose did have one positive effect: Ninety percent of the families kept their appointments—an astonishing rate. "You can't get around the fact," admitted Kreutzer, "that there is a great deal of anxiety that motivates parents to bring their children in."

The one-time appointments also turned up some important information about the overall health of McFarland's children. Of 1,744 children screened, 71 percent needed follow-up treatment. They were referred to other doctors, or more often, to the McFarland Community Health Center operating just down the block. Twenty-four percent of the kids showed signs of anemia, an indicator of malnutrition. Forty percent suffered from vision defects, and 36 percent needed dental care. Beyond these problems, McFarland's children did not seem less healthy than their peers in other parts of the country—at least when compared against the Health and Nutrition Examination Survey, a study conducted by the National Center for Health Studies every six years. This fact inclined state officials to waive their fears about some undetected, looming health catastrophe in McFarland.

"We had heard talk about premature puberty and bizarre skin rashes and all sorts of other health problems," said Dr. Kreutzer, "but we were never able to verify them. Now we are doing our best to reassure the community that at least some of their concerns aren't born out by the statistics gathered through the state agencies."

Beyond the reassurance gleaned from the state program, the children's visits to the doctor also produced a potential research tool to assist future investigations into McFarland's cancer cluster mystery. From all 1,744 children screened, the medical staff drew seven-to-ten cubic centimeters of blood, which was then frozen at minus-70 degrees and secured in freezers at the Department of Health Services' storage site in Berkeley. Researchers hoped that someday the technology might be developed to identify some genetic component of the cells that related to the town's childhood cancers. In the recent past, the Centers for Disease Control had used frozen blood specimens to decode Legionnaires' disease and to track the AIDS epidemic.

"The conventional wisdom has been that the cancer cluster could be a statistical fluke," said Kreutzer. "Now a statistical fluke under any circumstances is really just a nice way of saying: We don't know.

And the statistical fluke of today may have some obvious cause tomorrow."

When asked how long the blood specimens would last, Kreutzer replied: "Many, many years. Theoretically, indefinitely."

Once again, these words would return to haunt the Department of Health Services.

For the present, researchers and residents of McFarland would have to wait for the completion of the four-county study ordered by the state to determine if the childhood cancer rate within the entire region was elevated.

Then on September 13, 1989, the state health authorities startled the entire region with another entirely unexpected announcement. The officials had confirmed that one more small farming town in the Central Valley had emerged as a childhood cancer cluster.

Fifteen miles north of McFarland stood the town of Earlimart, population 6,000, an all-Hispanic community where six children had been diagnosed with cancer over ten years—about three times the expected rate.

The cancers had been tracked down by the United Farm Workers, then followed up by the state. State health authorities would not speculate about the cause of the cancers. The UFW did.

"I don't have a suspicion," declared Dr. Marion Moses, a longtime associate of the farm workers union. "I know what it is. These people live in a soup of chemicals down here. It's in the air, it's in the soil, it's in the water, it's everywhere."

Moses also made plain her disinterest in seeing Earlimart run through the same frustrating progression of scientific investigations that had absorbed so much time, energy, and money over the past five years in McFarland.

"We do not need any more studies," she insisted. "We do not need any more experts, or any more committees. All use of known and suspect carcinogenic pesticides in agriculture must be stopped and it must be stopped now."

Who Do You Trust?

The Riddles of Environmental Racism in Emelle, Alabama

Geiger, Alabama. The world's largest hazardous waste site sits between Geiger and Emelle.

The first threat to Kaye Kiker's life came when her husband Doug answered the telephone one hot, sultry evening in 1984.

"Keep your wife quiet," growled a stranger, his voice thick with the burly twang of rural Alabama. Then he threatened to burn down the Kiker's house, and hung up.

Over the years, similar threats would reach Kaye Kiker. Sometimes, it was difficult to tell exactly what action or public statement on Kaye's part had inspired the calls. But on this first evening, both Kaye and Doug knew exactly why the mysterious voice had threatened her life.

It was Kaye's recent appearance on television.

At the end of a long hot day, many of Kaye Kiker's neighbors in Sumter County, Alabama, settled down to watch the local evening news. But instead of a relaxing moment after dinner, they had found themselves gazing fixedly at an unimaginable sight: Hundreds of black people, assembled in the falling darkness, shielding candles in their cupped hands, were marching through the nearby town of Emelle, forming the largest protest to hit the region since the height of the civil rights movement 20 years before.

For many white and black Alabamans alike, recollections of the early 1960s remained painful and unsettling. Back then, the nightly television news broadcasts had focused upon the state's club-wielding police, their savage attack-dogs, and the water hoses unleashed against clots of unarmed demonstrators. These images had burned into the memories of many Alabamans. Now it appeared to be happening again.

Emelle was a very small, very poor, backwater Alabama hamlet, largely populated by African-Americans. But surprisingly, the television camera showed a smattering of whites among blacks at the protest; they were holding hands, singing protest songs, kneeling together on the ground as the Protestant ministers who organized the event called upon God to guide their actions. During the civil rights movement of the sixties, the unexpected collaboration of blacks and whites—the latter pouring into Alabama from points as distant as New York, Chicago, and San Francisco—had stiffened the resistance of Southern segregationists. Even now, the thought of this forbidden mix could enrage some white people. But this evening, most people watching the local television news realized that Northern whites could not have invaded the obscure small town of Emelle on such short notice. The whites, they knew, must be locals.

The news cameras panned along the line of marchers, their faces eerily lit by the candles cradled in both hands at their waists. At the head of the march, the Reverend Ben Chavis abruptly halted, turned to face the frozen glare of the giant television lights, and addressed the multitude.

As the camera drew in upon Reverend Chavis and the black people clustered around him, many TV viewers watching him from the more affluent, largely white small towns surrounding Emelle moved to the edge of their chairs. Standing alongside Reverend Chavis was Kaye Kiker—known throughout her community as an upright woman, a Sunday school teacher, president of the historical society, member of the homemaker's club. This pleasant, middle-aged, conventional housewife, who described herself as "the type of woman who went to church, didn't question authority, and was kind and hospitable like my Southern mama taught me to be," seemed the least likely person imaginable to join the procession of protestors.

To many people in the area, the sight of Kaye Kiker at the front of the march was disorienting, even shocking. But her presence was only the first surprise that evening for the television viewers of Sumter County.

At the rear of the protest trailed a small, elderly white woman, clutching a candle close to her breast, singing "We Shall Overcome."

It was Kaye Kiker's mother.

■   ■   ■

Deserted by cotton and bypassed by industry, Emelle, Alabama, appeared to be just another sleepy Southern town—a bare pinprick of population, holding steady at 626, cloistered from passersby amid its drooping cypress and willow trees, separated from the sprawling countryside by a borderline of swampy furrows. Yet two dubious and intimately entangled distinctions had recently elevated Emelle to national prominence.

Emelle was home to the world's largest toxic waste dump, renowned as "the Cadillac of Landfills," and winning sufficient acclaim for its owner-operator, Chemical Waste Management, Inc., that even *National Geographic* celebrated its state-of-the-art design in a lavish, full-color pictorial. Consequently, this small town, with a 79 percent African-American population, had also become the symbolic center of a mounting controversy regarding "environmental racism"—an ugly concept based on the fact that the majority of the nation's toxic chemical wastes are dumped in African-American, Hispanic, and Native American communities.

The recognition of this pattern did not begin with the Emelle dump. Rather, the first vigorous interjection of minority concerns into the contemporary environmental movement started in the early 1980s, hundreds of miles away, in North Carolina.

In 1982, North Carolina governor James Hunt, in conjunction with the EPA, announced that 30 thousand gallons of toxic PCBs would soon be buried along the northern edge of the state, in Warren County. This area had a higher percentage of blacks than any other county in North Carolina—and only a scant accumulation of traditional white, middle-class environmentalists who could be expected to object. And yet, the governor could not have selected a more volatile site.

"I think it's fair to say," recalled Dr. Ben Chavis, executive director of the United Church of Christ's Commission on Racial Justice, "that the governor of North Carolina made a mistake when he assumed the residents of Warren County would allow the dumping of the state's PCBs in their neighborhood without protesting it."

Yet even Chavis, a seasoned civil rights activist who had received a stiff prison sentence as a member of the Wilmington Ten, did not immediately grasp the national implications of Warren County's response. The decision to ladle PCBs upon the predominantly black county hit a nerve among poor, rural people who had previously been too concerned about jobs, housing, and poverty to worry about the environment. Yet when the trucks rolled into their county to deliver their toxic loads, hundreds of residents blockaded the road. State police arrested more than five hundred protestors. On his way to address the crowd, Chavis was pulled over by a patrolman and arrested for driving too slowly. "Their way of keeping me from speaking at the rally," he said, "was to put me in jail."

Stunned by the size of the first nationally recognized environmental protest within a black community, the press descended upon Warren County, speculating that the overwhelmingly white environmental movement might have finally achieved some alliance with minority constituencies. In truth, the national environmental organizations contributed nothing to the Warren County uproar. Blacks in the region had organized themselves, according to their own concerns—and they drew inspiration not from the mainstream environmental movement, but rather from their own neighbors' esteemed role in the civil rights movement. The lunch counter sit-ins in Greensboro, Durham, Raleigh, and other North Carolina cities remained an immense source of regional pride. Now, two decades later, the people of Warren County viewed the instant transformation of their community into a toxic dumping ground as an extension of the same disregard and contempt for black lives that allowed poverty, crime, poor schools, inadequate housing, and disease to flourish at far greater rates than in white communities. From the beginning, the Warren County demonstrators—like their white working-class counterparts in other contaminated communities—cared less about environmental purity than environmental justice.

When the demonstrations concluded, the press departed Warren County, but evidence continued to mount of a nascent environmentalist ethic growing within the traditional black social agenda. Black activists throughout the United States contacted the leaders of the Warren County demonstration, revealing that their communities also had been slated as hazardous waste sites. Even in predominantly white

regions, they complained, the dumps seemed to always end up in
minority neighborhoods.

For the next three years, Chavis and the Commission for Racial
Justice collected stories and data indicating a profound national pref-
erence for African-American, Hispanic, and Native American communities as storehouses for toxic wastes. In addition to the documented persistence of discrimination in employment, housing, and
loan practices, they began to formulate a notion of environmental
inequities predicated on race. Prior to their speculations, black leaders had never even raised the issue. "The most significant thing about
Warren County," said Chavis, "is that the idea of environmental
racism, and the need for environmental justice, emerged out of the
struggle there."

In 1986, the Commission for Racial Justice decided to quantify
the information about environmental racism that they had heard over
and over again, but had not verified. In a national survey of the
nation's licensed hazardous waste sites, plus some 20,000 abandoned
or unregulated toxic dumps, the commission found "a striking relationship between the location of commercial hazardous waste facilities and race."

While only 26 percent of the country's population was African-
American or Hispanic, 60 percent of the largest toxic waste facilities
ended up in African-American or Hispanic communities. Three out
of five African-Americans and Hispanics lived near "uncontrolled"
toxic dump sites. Regions dotted with hazardous waste problems
revealed minority populations two-to-three times higher than relatively unblemished parts of the country—with cities containing large
African-American communities, such as St. Louis, Houston, Cleveland, Chicago, Atlanta, and Memphis, hosting the greatest number of
urban dumps. About half of all Native Americans also lived near toxic
waste sites. And in recent years, reservation land had been frantically
courted by private entrepreneurs to house chemical and nuclear
wastes, since tribal law superceded more stringent federal and state
disposal regulations.

After compiling its findings, the commission employed the New
York research firm Public Data Access, which reviewed and confirmed
the study's statistical analysis. To Chavis and his peers, the report
clearly indicated that the chief factor in determining the location of
toxic dumps was race. Industry and government critics countered that
it was not race, but economics: Waste disposal companies sought the
cheapest available land for their enterprises, which tended to be
inhabited by poor people; among the poor, minorities were unfortunately—but coincidentally—overrepresented.

Yet Chavis argued that the guiding factor of race was blatantly unveiled in the Deep South and Appalachia, where poor whites abounded—and where the majority of hazardous waste dumps still ended up in black communities. "We tested for poverty and we tested for race," maintained Chavis. "Race was the most predominant factor in where the facilities were located, not poverty. In this country, what makes property cheap is largely determined by the race of the people who live there."

Armed with this incendiary analysis of the statistics, Chavis coined the term, "environmental racism," and drew renewed interest from the press and government policymakers. Sensing the opportunity to illustrate their point with banner headlines, the leadership of the United Church of Christ, including Chavis, Dr. Charles Cobb, and Charles Lee, stormed into Emelle, Alabama, intending to demonstrate that there was a national problem, symbolized by this one tiny African-American community containing the world's largest toxic waste dump.

Yet Chavis and crew had not calculated the crucial differences between Warren County, North Carolina, and Sumter County, Alabama.

Unlike Warren County, political power in this remote corner of the Deep South still lingered in white hands. Until 1984, white elected officials controlled the County Commission, despite the region's substantial black majority. It was not until 1985 that Sumter County sent its first black representative to the state legislature. Sumter County was "a classic case of apartheid, American-style," according to Professor Robert Bullard, a sociologist at the University of California at Riverside, who studied the pattern of toxic waste disposal in black communities throughout the South. "You've got a small white minority that's been running the county for 250 years," said Bullard. "And still is."

Bullard termed Emelle's politics a "plantation power arrangement," but its parochial strugggles also indicated a larger conflict involving the state's long-standing tribulations over equity and race. Thanks largely to the federal Voting Rights Act of 1965, Alabama had elected over seven hundred black officials to all levels of government—more than any other state in the nation. However, the deeper sources of political power—money, ownership of land, and access to state and federal decision makers—still eluded leaders in most black communities. In a sense, the basic questions of self-determination first posed during Reconstruction had not been fully answered in places like Sumter County. The conflict over the Emelle dump drew into focus the naked mistrust and disaffection between blacks and whites that still characterized American life throughout the entire nation after three centuries.

When people living near Emelle flipped on their television sets and found Kaye Kiker and her mother marching alongside Chavis and a contingent of black demonstrators from Sumter County, their disbelief made perfect sense. The Emelle demonstration against the Chemwaste dump marked the first time that blacks and whites in Sumter County had joined together in a public protest over any political issue.

Yet beyond the dramatic images broadcast by the television news cameras, the people watching television had no idea what was really happening behind the scenes.

Although local black leaders warmly greeted Chavis, Cobb, and Lee when they first arrived in town, they also took pains to impress upon their guests the absence of a solid black political opposition in Sumter County—unlike the visitors' activist home territory in North Carolina. "We kept telling them that this really wasn't going to work out so well here," recalled John Zippert, editor of the *Greene County Democrat*, a black newspaper published in the adjacent county. "There was a lot more work that needed to be done to organize and inform this community. But the candlelight prayer vigil and demonstration was what they wanted to do—and they got their day in court, so to speak."

To help organize the demonstration, Zippert joined with Wendell Paris, a fellow member of the Federation of Southern Cooperatives, a black agricultural organization. Together they scoured the county for residents concerned about the dump's health and safety liabilites. Paris also approached members of the predominantly white Alabamans for a Cleaner Environment—Kaye Kiker's group.

When the night came for the demonstration, the North Carolinians—Chavis, Cobb, and Lee—joined with Zippert, Paris, and about one hundred local black protestors, as well as Kaye Kiker, her mother, and a smattering of other white faces. The group lit their candles, cradled the flames against the wind, and marched to the front gates of Chemical Waste Management in Emelle. They stood silent in prayer for a moment.

Suddenly the plant's floodlights drenched the crowd in a vast radiant beam. The demonstrators' flickering candles faded against the torrent of intense light. Workers from the dump poured out around the gates.

"Don't jeopardize our jobs!" one man shouted at the demonstrators.

"We support Chemwaste!" hollered another worker.

The silent prayer vigil concluded in fulsome strife, as workers swarmed around demonstrators, barking their objections. "We're talking past each other," John Zippert cried out from the crowd. Nobody seemed to hear. "My name is John Zippert," he yelled. "My phone number is listed in the phone book. MY HOME PHONE NUMBER! Call me! Talk to me!"

From the Chemwaste offices, Gordan Kenna watched the demonstrators disappear under the waves of counterdemonstrators. Kenna had taken over the job as Chemwaste's community relations director only four weeks earlier. Now he stared directly at a potential public relations catastrophe.

Corraling Zippert at the edge of the crowd, Kenna quickly improvised a meeting in the cafeteria between the demonstrators and workers, insisting upon the need to cool off both sides. "They brought us into a big room with some 250 employees," recalled Zippert. "Reverend Chavis and Reverend Cobb faced the workers and said a prayer. They made a statement of their concern about facilities like this being put in minority communities. The workers immediately started shouting. *What right do you have to come here? You guys are here to take away our jobs.*' Reverend Chavis and Reverend Cobb had come to Emelle to make a point that was much larger than the workers' personal concerns. They wanted to make a moral witness against environmental racism. They wound up at the wrong end of a workers' pep rally organized by Chemwaste."

"I guess they got what they came for," sighed Zippert. "And then they left again. And they never, quite honestly, have really helped us do something here."

- - -

The protestors who clashed with workers at the gates of Emelle's huge dump feared the health hazards posed by toxic wastes. A few of them had heard about Love Canal and Times Beach—but more importantly, they knew about the less celebrated, but more localized toxic disasters, such as Triana, Alabama, once characterized in *National Wildlife* magazine as the "unhealthiest town in America." Located in northern Alabama, the residents of the all-black town registered in their blood tests some of the highest levels of DDT ever recorded, due to eating fish from a creek contaminated by a local chemical manufacturer.

Beyond their knowledge of toxic disasters, the people of Emelle were also implicated in a public health catastrophe unrelated, but disturbingly coincident with the spread of chemical dumping in

John Zippert.

minority communities. Throughout the 1980s, the health status of African-Americans had plunged to staggeringly low levels. Diseases considered on the verge of eradication only five years earlier were staging a huge return among the inner-city and rural poor: Blacks, Hispanics, and other minorities faced skyrocketing rates of tuberculosis, hepatitis A, syphilis, gonorrhea, measles, mumps, whooping cough, complicated ear infections, and AIDS. In the *International Journal of Epidemiology*, scientists at the Washington, D.C., Commission on Public Health estimated that blacks in the nation's capital were four times more likely than whites to die before the age of 65 from heart disease, asthma, pneumonia, and some cancers. Death rates for blacks in Washington had climbed steadily since 1982; for whites, the rates had not changed.

"We looked at conditions that people shouldn't be dying from in the prime of life," explained Dr. Eugene Schwartz, one of the study's authors. "And remember, death is a very crude measure. It's the tip of an iceberg, the end of a long chain of illness and disability that we don't have a handle on."

According to the *New England Journal of Medicine*, a black man living in Harlem was less likely to reach the age of 65 than a poor man in Bangladesh because of his community's high rates of disease. "We're using Band-Aids and things are getting worse rather than better," said Dr. Harold P. Freeman, chief of surgery at Harlem Hospital. "You have to see disease in the context in which it occurs: poor education, unemployment, homelessness, hopelessness."

This context also affected the way in which many black people in Emelle viewed the Chemwaste dump. The national decline in health for African-Americans, coupled with the local siting of the world's largest toxic waste facility, offered an irresistible symbol of simultaneous assault. The relationship between the wide variety of escalating health problems, and the preponderance of dump sites and incinerators in minority communities, did not have to be causal to rouse tremendous feeling; its convergence proved sufficient.

Even more important was the community's basic skepticism about government—a sensibility evidenced with sufficient strength in all-white, working-class contaminated towns, but amplified monstrously by the complexities of Emelle's divisive history of racial politics. The continuing controversies over justice, political power, public health, and economic development were nearly inseparable since many of the same local adversaries had faced each other on these issues for three consecutive decades.

The most pointed example of Sumter County's polarization of black and white interests could be located in the persistent rivalry between Wendell Paris, one of the local black organizers of the ill-fated Emelle protest, and Drayton Pruitt, the resident prime example of white Southern gentry.

Wendell Paris grew up in a family inclined to resist the status quo. In 1941, Paris' father served as Sumter County's first black agricultural agent, distributing federal aid checks to local farmers. "Dad was a brave man," remembered Paris. "He drove a lot and needed a good car, but he wore a chauffeur's cap because whites would stop a black man driving a car that looked too good. So he'd tell them, 'I'm the driver for Mrs. George Paris'—and they'd let him go." Wendell's father never mentioned that Mrs. George Paris was his wife. In case of a more complicated confrontation, he carried a sawed-off shotgun under the seat and a rifle in the trunk.

George Paris also distinguished himself as one of Sumter County's first registered black voters, braving the hurdles of poll taxes and irrelevant tests that had been erected to exclude blacks from the electorate. "The judge asked him to recite the U.S. Constitution," recalled Paris. "So he puckered up and started in on the Preamble. *We the people.* . . . And the judge stopped it. 'It's clear he knows the thing,' the judge said, 'and we don't.' So Dad was registered."

While George Paris was reciting the Preamble to the Constitution to secure his right to vote, Drayton Pruitt's father was occupying his seat in the Alabama state legislature, which he held for 30 years—and which as a body had blocked black voter registration for nearly a century. If the Paris family was an avowed adversary of the status quo, the Pruitts were its incarnation. As the civil rights movement of the early 1960s penetrated the Deep South, Wendell Paris naturally drifted towards the center of the storm. In 1965, he marched in Selma with Martin Luther King, returned to Sumter County to organize the Alabama Self-Help Association, and finally established a local branch of the Federation of Southern Cooperatives, an organization attempting to counter the declining black ownership of farmland. Throughout the sixties and seventies, Paris cultivated a reputation as Sumter County's preeminent civil rights radical, admired and trusted by blacks, often feared by whites.

On the other side of town, young Drayton Pruitt rose quickly through the local political and business establishment. He was elected mayor of Livingston, the county seat, and also served as the Sumter County district attorney, city attorney for the nearby towns of Epes and Geiger, and attorney for the County Commission. He sat on the boards of directors for the region's major banks. And as Governor George Wallace's west Alabama political representative, he won confirmation from the state senate to sit on the Alabama Ethics Commission.

By the 1970s, the younger members of the Paris and Pruitt families found themselves in direct conflict. The federal Voting Rights Act of 1965 had finally been implemented in rural Alabama, and numerous Sumter County blacks registered to vote—only to be promptly thrown off their rented farms.

"Some 250 families were evicted overnight," recalled Paris. "These folks had grown cotton all their lives on those small farms." Through the Federation of Southern Cooperatives, Wendell Paris worked with several civil rights groups to help local black families purchase new farms. When the fight over a particular parcel of land went to court, the lawyer opposing Paris and the black farmers' side was Drayton Pruitt.

"It was just a business thing," recalled Pruitt. "Unfortunately, the people at the Federation were so emotionally charged about it that they took things personally."

This conflict primed both sides for the larger battle that would soon follow.

Throughout the 1970s, Governor George Wallace's son-in-law, Jim Parsons, along with two other partners, started buying up large plots of land near Emelle. They formed a corporation called Resource Industries in 1977, hired Pruitt as their legal representative in the county, and soon unleashed a squad of heavy mechanized equipment to scoop out enormous pits in the ground to prepare for their new business.

Paris first heard about this venture when he picked up his copy of the *Sumter County Record* on May 25, 1977, and scanned the promising headline: "Unique New Industry Coming: New Use for Selma Chalk to Create Jobs." Understanding that political power hinged on economic clout, Paris knew that new jobs could substantially improve the situation of Sumter County's blacks in many ways. The Selma chalk mentioned in the headline referred to a white, brittle, nonporous rock that ran hundreds of feet deep, covering an expanse from central Mississippi into central Alabama. Many Sumter County residents presumed the new industry coming to town would be a brick factory.

But Sumter County's newest arrival was destined to be Chemical Waste Management, a subsidiary of Waste Management, Inc., the nation's largest corporate waste handler. The local group had sold their entire operation and land holdings to Chemical Waste Management for an enormous fee, stock in the company, and 20 years worth of royalties on each barrel of toxic waste buried in Emelle. Chemwaste wanted to locate its new dump in Sumter County, taking advantage of the EPA mandate to store the nation's toxic overload in 36 privately operated regional centers, rather than thousands of smaller facilities run by county governments. This plan was based on the reasonable assumption that a smaller number of controlled sites would be easier to monitor and regulate. The massive quantities of waste stored in these mega-dumps would also underwrite more expensive, and presumably safer, disposal technologies. Since the Emelle site was among a select handful of sites that had actually moved from the merely speculative stages into operation, hazardous materials from 46 states were going to be trucked into the small Alabama hamlet for treatment and burial.

When Wendell Paris discovered that the new industry in Emelle would not be bricks hewn from Selma chalk, but rather toxic wastes contained by this dense, purportedly impermeable substance, his enthusiasm faded. And when he discovered that the local front man was Drayton Pruitt, Paris became an unswervable opponent to the dump.

In Paris' view, the entire county had been duped by Pruitt and his partners, whom he insists played down the presence of hazardous materials in their new enterprise. "There was absolute white control of the county in 1977," said Paris. "You had commissioners who had been in office for 30 years. And so here's Drayton, holding all of his positions in government, and he joins up with George Wallace's son-in-law. Sumter County was ideal for them in the 1970s because unlike nearby counties, there wasn't an informed, active black population in the political arena. We didn't have any idea they were planning to bring in hazardous chemicals.''

In Pruitt's opinion, Paris understood the businessmen's plans from the start. "Wendell is a smart man," he said. "I don't think he'll ever convince me he didn't know what was going on here."

According to Paris, his first understanding of the dump's operation arrived as a shock. One afternoon in 1979, Paris was working in his office at the Federation of Southern Cooperatives in Epes when some 50 Chemwaste workers showed up at the door. They were covered in mud, muck, and sweat, and they were angry. " 'We had to walk out of there,' " Paris recalled them saying, " 'because of the hazardous working conditions.' " As Paris tells the story, one of the workers had driven a bulldozer down into the bottom of a pit, carrying barrels of chemical waste. "The bulldozer's brakes were shot," said Paris. "About halfway down, the thing slipped in the Selma chalk and skidded sideways all the way to the bottom of the pit. The managers got into a four-wheel drive and went down after him. 'Now get back on the machine and go back to work,' they told him. The driver refused. They told him, 'Get back on it—or go!' The driver walked off, along with about 50 other workers who had watched the scene."

Wendell Paris in Emelle.

Drayton Pruitt in his office in Livingston.

One of the workers had brought with him a tag taken from a barrel he had buried. The tag included a telephone number to call for information. "I called the number," remembered Paris, "and I got some kind of oceanography place. 'What's Emelle, Alabama, got to do with oceanography?' I asked. The guy on the other end of the line said, 'They're burying our toxic wastes there. Didn't you know that?' No, I didn't know that," affirmed Paris.

Paris and the Federation immediately became outspoken advocates for employee safety at the dump. But Paris nursed another fear that extended far beyond the health of individual workers. Below the layer of Selma chalk lay the aquifer that provided drinking water for a large portion of western Alabama. "Because of what we learned from the workers," said Paris, "we saw there was a direct threat to our whole community."

To Wendell Paris, and his associate at the Federation, John Zippert, the chemical waste dump qualified as the latest insult and injury to the black residents of Sumter County.

"These dangerous facilities," John Zippert insisted, "are put in communities of people considered to be marginal, not as important or valuable as white people. If you consider people to be lesser human beings, it doesn't matter if you dump this shit on them. And if you look at the national pattern, we dump our hazardous wastes in communities of people that we consider to be more expendable—which is a racist view of the world.

"We didn't understand what the place was about until it was already there," insisted Zippert. "And the first people affected were black people who worked there. Drayton Pruitt took advantage of this county, coming into a community where you don't have people with resources. And he didn't tell them what was going on."

To Drayton Pruitt, and much of the white power structure that had welcomed Chemwaste to Emelle, the protests headed by Paris, Zippert, and the Federation were merely a sign of self-defeating intransigence among would-be black leaders who had for years stalled an economic renaissance in the county.

"There were no public disclosure requirements in the law at that time," maintained Pruitt. "The hazardous waste plant was an industrial development tool for a poor county. The majority of people in this county, I believe, saw it as the beginning of locating some good industries here. And to be honest with you, if it hadn't been for all the protests and the media hype, that's just what would have happened."

-   -   -

We met Gordan Kenna, the director of community relations for the Chemwaste dump in Emelle, at the company's front office. Kenna planned to spend the afternoon with John Zippert and some other representatives from the Federation of Southern Cooperatives. Apart from the demonstration held several years before, Zippert had never visited the dump site. Kenna hoped to assuage his fears and establish a relationship that would reach the leading skeptics within Sumter County's black community.

While we waited for the Federation members, Kenna explained why Chemwaste now concerned itself with the good opinion of John Zippert, Wendell Paris, and other local people who had fervently opposed the dump for many years.

"Chemical Waste started out with support in Sumter County, but it came from white people at the top of the power structure," admitted Kenna. "Now the balance of power has changed radically. Yet we can't just instantly shift with the times. It's a delicate balance. We've got to attract new support, without alienating the old support."

When Emelle's toxic dump was first sited in 1977, the governance of Sumter County rested in the hands of an all-white, three-man County Commission. Challenges to the power structure rarely occurred, and in the case of the dump, neither black nor white residents initially raised objections. The year before, when the local industrial board sponsored a public meeting to explain the purpose of the dump, the term "hazardous waste" had not yet become a household word. The revelations of Love Canal, the nation's first widely publicized toxic disaster, would not make national headlines for another two years.

By 1991, the situation had changed drastically. Even in remote Sumter County, people knew about hazardous wastes. And black people were no longer cut out of the local political process. The County Commission had expanded to six members—with four blacks holding the majority. Chemwaste could ignore these changes only at its own peril.

"Chemical Waste bought out this facility in 1978," explained Kenna, "but it wasn't until 1980 that the EPA had any legal responsibility for hazardous wastes. Back then, if you didn't dispose of hazardous wastes properly, you were basically just violating local litter laws. It wasn't like you'd have a problem with the feds or something. Then Superfund became law in 1980. But in January of 1981, Ronald Reagan became president, and he appointed Anne Gorsuch Burford to head EPA, Rita Lavelle to head Superfund, and James Watt to head the Interior. So who was watching the environment? Nobody. Until, frankly, the American people said, 'Wait a minute!' I think that's what democracy is generally about," insisted Kenna. "People saying, 'Wait a minute!' "

If Gordan Kenna sometimes sounded more like a Greenpeace environmentalist than a public relations officer for the nation's largest hazardous waste firm, the difference could be partially explained by a set of career steps that defied the conventional progress into industrial flackdom. From 1975 to 1977, Kenna served as an aide and senior policy advisor to Mayor Maynard Jackson of Atlanta, the first black mayor of a major Southern city. Kenna worked with the first Jackson administration to ensure that minority construction firms would participate in city projects, including the massive expansion of the Atlanta airport.

After leaving city government, Kenna continued to focus on minority job development with the Southern Regional Council, a policy group formed early in the century to edge the South towards greater racial and economic justice. Kenna once again concentrated on gains in industrial jobs, studying company employment records to fight discrimination against women and minorities.

In 1979, Kenna left the South to join the Environmental Protection Agency. During his tenure at EPA, Congress established the first regulations governing hazardous waste disposal and affirmed the policy directing the nation's toxic load to a select number of large, iso-lated disposal sites, rather than thousands of smaller dumps.

"My role in the EPA," explained Kenna, "was to explain to peo-ple in affected communities what EPA was doing—that we were pick-ing up the chemicals and taking them away someplace so that folks would be rid of them." The "someplace" to which the chemical wastes were destined turned out to be towns like Emelle, Alabama.

"I had experience in dealing with hazardous wastes," summa-rized Kenna, "and I've had a long interest in racial and social justice issues. So when I took this job at Chemical Waste Management, it was a natural fit for me."

Kenna presented a textbook example of the ideal public relations officer. Affable, but not unctuous, open although clearly in control, the 44-year-old Kenna exuded the smooth sincerity of a man other people *wanted* to trust. And his impeccable resume only strengthened the image.

When John Zippert finally arrived at the Chemical Waste Man-agement office, accompanied by four other Federation members, Kenna smiled, extended his hand, and offered each person a chair around the large, polished-wood conference table.

Zippert resisted the charm.

"To be honest," Zippert curtly informed Kenna, "*we have a lot of concerns about Chemwaste.* You said you've got better processes now of handling the waste, so we came out to see what's going on before we mounted our criticisms. But we will highlight the question of why these sites are in minority communities. And we want to know what protection the employees have, and what protection the community has. We've been told for years that this place is here because of the geology of the Selma chalk—and we'd like to get a better understand-ing of the validity of those claims. And the people involved here: How many really come from Sumter County? And we want to know the black-to-white ratio." Zippert had barely taken his first breath.

"The people in this community feel like they're not being rec-ognized," summed up Cleo Askew, a large black man with a soft, low voice who was a member of Zippert's delegation.

"Those are real issues," agreed Kenna, a professional. He nod-ded promisingly to both men.

"We are *sitting* here in your office," complained Zippert, clearly growing more uncomfortable, searching in vain for something to push up against, "and we really can't *tell* who is *on our side.* Is the EPA *on our side?* Is the Alabama Department of Environmental Manage-ment *on our side?* Who exactly is *on the side of the community* and of the people who *work* here—the people most vulnerable to potential health damage from this facility?"

Kenna assured Zippert that all plant employees received annual health tests. Doctors had found no significant effects of hazardous waste exposures among them.

"In the end," declared Kenna, "it comes down to the question: *Who do you trust?*" He spoke directly to Zippert. "Is the company worthy of trust? Is the regulatory agency worthy of trust? Are the community representatives sufficiently objective so that they can be trusted both by the community and the company? We are running a business here," he reminded everybody. "But we don't give people that warm, fuzzy feeling, like a company selling hamburgers or soda pop."

Kenna rose from his chair, shrugged off his sports coat, and rolled his white shirtsleeves halfway up his forearms. "I'd like to find a happy medium here," he told Zippert, "and maybe today's meeting is a good start. Let's take our tour of the facility and show you what this company's all about."

Outside the office, we piled into a large, air-conditioned tour van, along with Kenna, Zippert, Askew, and other Federation supporters. As the van rumbled across the steaming tarmac towards the plant's vast pits where the hazardous wastes were buried, Kenna reviewed the virtues of Selma chalk, stressing its impermeability. In addition to the limestone running hundreds of feet beneath the waste pits, the company also coated all disposal areas with heavy plastic liners. Constant monitoring of 8 deep groundwater wells and 60 shallow wells ensured that the operators would know immediately if contaminants managed to reach the aquifer.

We stared out the mini-van's windows, squinting into the harsh sunlight, trying to make sense out of what we could and could not see. The mini-van halted at the top of a grassy knoll. Everybody climbed out. "At every meeting with environmentalists that we attend," objected Cleo Askew, "we're told that this facility is going to leak, that this facility could cause water problems for the entire county."

From the top of the knoll, as far as we could see, green, grassy hills rolled off into the horizon. Between the hillsides stood numerous metal huts, containing chemical wastes waiting to be buried. Somewhat closer, heavy machinery screeched and grumbled, scooping out deep burrows in the ground. After the digging was completed, the black metal barrels of chemical wastes were slowly, *gingerly*, lowered into the pits, and later covered with dirt. Over time, grass and flowers would sprout up over the burial sites, erasing the industrial scars. Of the 2,700 acres owned by Chemwaste in the Emelle area, only 350 acres were currently used for waste disposal. From our view on the grassy knoll, the Chemwaste operation offered visceral assurance of order and continuity.

"It's far more likely," Kenna assured us, wiping the sweat from his forehead, then glancing up at the sky, "that this sun will burn itself out than that we will cause water contamination problems here."

Yet for all the certainty implied by our view from the grassy knoll, John Zippert could not—would not—shake his skepticism about Waste Management, Inc. Zippert had tracked down the parent company's trail of costly environmental violations, and much of what he had learned ran contrary to the image of trust and accountability emphasized by Gordan Kenna.

According to a survey conducted by the Environmental Research Foundation, Waste Management had been repeatedly sued for millions of dollars by the EPA and various state regulatory agencies for numerous violations of standard safety practices at its waste disposal sites throughout the country. One of the most common violations involved inadequacies in monitoring the groundwater near the plants for chemical contamination.

In California, the EPA collected $4 million in fines from Waste Management for its faulty groundwater monitoring system. The company explained that its officials had some "differences of opinion" with the EPA regarding the proper monitoring procedure, but evidently these differences could not be resolved. At the Waste Management facility in Vickery, Ohio, the EPA once again leveled fines—this time for $10.5 million—for actually contaminating the groundwater. The EPA had also fined the company at the Emelle site for $1 million over numerous safety violations, including problems with groundwater testing. In fact, over the years, according to an account in the *Wall Street Journal*, Waste Management had accrued so many fines and violations that it was forced to take the extraordinary measure of paying $11.4 million back to its own stockholders for failure to disclose liability from "investigations, complaints, enforcement actions and other suits" against the company.

Beyond these regulatory entanglements, the public notoriety of several Waste Management associates also had stirred John Zippert's suspicions. In 1987, a lobbyist for the company pleaded guilty to bribing a Chicago alderman in return for favors in siting a hazardous waste facility in the area. In Niagara Falls, New York, plant officials had been cited for filtering water samples before testing them—thus violating an EPA protocol and potentially masking the level of contaminants found in the water. In San Diego, the county district attorney issued a scathing report to the board of supervisors (who were then considering a contract with the firm) stating that "Waste Management engages in practices designed to gain undue influence over government officials. . . . These practices suggest an unseemly effort by Waste Management to manipulate local government for its own business ends. If unchecked, these practices . . . may have a corrupting impact on local government and lead to decisions unsuitable to the best interests of the public."

In sum, the history of the company, as John Zippert assessed and understood it, did not make him feel very trustful at all. Even from the reassuring perspective of the grassy knoll pitched high above the clockwork operations of the Emelle dump site, Zippert viewed the future with disbelief and anxiety.

And the future itself raised another set of troubling questions: What happened to oversight at the dump if Chemwaste went out of business? What happened in 30 years, when their statutory obligations to oversee the dump's safety ran out? Nobody could really answer these questions because the situation was unprecedented: This amount of toxic material had never been plowed into one place before. In the end, Zippert was left imagining Emelle as the centerpiece of a vast national sacrifice zone—surrounded, inevitably, by black people who could not afford to move away.

Kenna led us back into the mini-van, and we headed to the laboratory where Chemwaste conducted its testing and analysis of arriving wastes, and the dump's own groundwater samples. Within a few minutes, we had assembled in a narrow corridor, facing a half-wall of plate-glass windows sealing off dozens of workstations. Behind the glass panels, we watched scores of white-coated technicians laboring busily at their gas chromatographs, atomic absorption spectrometers, and super-critical fluid chromatographs. Laboratory assistants hovered over veinous networks of glass tubing filled with water, feeding their samples into enormous white-enameled machines that split the fluids into hundreds of chemical components, analyzed them, and spit out their findings on reams of computer paper.

"I think we are a legitimate business that performs a very necessary, very important service," declared Kenna, beaming with pride as he raised one arm in a half-moon arc, as if to encompass the mass of testing equipment and busy technicians. "We ought to be accorded, in Aretha Franklin's words, 'a little respect.' And I don't think we get it."

John Zippert and Cleo Askew stared at the maze of incomprehensible technical equipment whose performance they had come to evaluate. They didn't utter a sound: completely out of their element, overwhelmed. Then Zippert shattered the silence.

"Gordan," he announced, "I think you've hired people from Mississippi to work here, so local people won't raise questions about the plant."

Kenna gazed at Zippert as if he were speaking an indecipherable foreign language.

"John," Kenna finally replied quietly, "that's entirely out of bounds."

"I'd like to see your records," shot back Zippert. "I want to find out how many people who work here actually live in Sumter County, and how many black people work here, and how many come across the state line from Mississippi, and how the division of jobs is split between blacks and whites. And I want Chemwaste to be taxed to provide funds for this community to hire an independent chemical consultant to come here"—Zippert surveyed the small city of complicated equipment surrounding him—"*and watch over just what is going on here.*"

Kenna informed him that the company did not keep records of its employees' county or state of residence, or their racial backgrounds.

"I've been lots of places," said Zippert, "and when people tell me they 'don't have the records,' it doesn't make me feel more trustful."

Kenna appeared stunned by the level of suspicion and disbelief confronting him. Zippert looked equally amazed that he was being asked to take the good intentions and sound practices of the waste facility—the maze of people and equipment in front of him—on faith. In silence, we all walked back out into the blazing afternoon sun, piled into the mini-van, and returned to the office.

Zippert promised to soon forward a letter outlining his thoughts on the day's meeting. Kenna agreed with more enthusiasm than conviction that a future meeting might be helpful. About a month later, Zippert mailed a note on Federation letterhead thanking Kenna for the tour of the facility, and suggesting the initiation of an "independent citizens' review board to monitor safety of your activities in Emelle." He also once again requested data on the race, county residence, and job positions of Chemwaste employees. One year later, Zippert still had not received a response to his letter. Kenna had left his job at Chemical Waste Management. The new community relations officer chose not to continue talking with Zippert and the Federation, and the situation returned to a quiet state of unarticulated animosity.

Cleo Askew confronted Gordan Kenna in the hallway of the laboratory.

Later that week, Chemwaste layed off 53 employees from the Emelle plant. The state legislature had recently increased taxes on each barrel of toxic wastes hauled into Alabama from out of state. Calculating a revenue loss, Chemwaste scaled back the workforce.

"It's economic blackmail," complained Kaye Kiker, whose local organization, ACE (Alabamans for a Cleaner Environment) claimed credit for pressuring the state legislature into increasing the tax.

Kiker was right that the layoffs had focused attention on Chemwaste's economic role in Sumter County. In addition to the 53 employees hoping to eventually return to work, Chemwaste also kept on file applications from another three thousand people coveting one of the Emelle facility's four hundred full-time jobs. Chemwaste calculated that its largest facility maintained an annual payroll of $11 million, spent over $20 million in local purchases, and paid almost $4 million in county taxes. In this part of the country—where, as Gordan Kenna put it, "you notice a twenty dollar bill"—Chemwaste often supplied the critical measure of financial support that kept municipal services afloat.

Chemwaste had built Emelle's new city hall, at a cost of $350,000. Corporate donations supported the local Boy Scouts and sports teams. The company's county taxes, accrued at a rate of $5 per ton of wastes

shipped into Emelle, underwrote police services, the water department, the rescue squad, the County Commission, and Industrial
Board. Since most of Sumter County's white students had fled the
public school system for private schools, the largely black educational
system depended heavily on the industrial tax base borne by the waste
disposal company.

According to Wendell Paris, Chemwaste was "turning Sumter
County into the pay toilet of America, and local residents into hazardous waste junkies." But when Paris was hired by the County Commission, he also was advised that part of his salary would come from
Chemwaste funds. Even John Zippert confessed that the Federation
of Southern Cooperatives' youth training program depended on a
$400-per-month subsidy derived from Chemwaste taxes.

"Chemwaste might just have the best landfill in America here,"
admitted Zippert, "but with the amount of money involved, we should
have a healthy amount of skepticism about what is going on. Most of
the black leaders are in political positions that receive money from
Chemwaste. They're all in a difficult position to speak out against
Chemwaste. And they are legitimately concerned about anyone who
just irrationally says, 'Close the place down.' Look," emphasized Zippert, "there are some groups in Sumter County who simply want this
place shut down. That's not our position."

And yet closing down the dump was exactly what ACE, Kaye Kiker's group, wanted to do.

Early one evening, we drove about a half hour outside of Emelle
to meet with ACE activists in a sprawling old plantation home, located
on 40 acres of farmland, and owned by Charles and Linda Munoz.
The Munoz family had returned to the Deep South after years spent
in northern cities, and their repatriation had proved difficult.

Two years after moving back to Sumter County, Charles Munoz,
a cellist who had trained in Europe, joined the board of directors for
the County Arts Commission. "We had a glorious budget," he
recalled. But he soon discovered that most of the commission's money
came from Chemical Waste Management, located in Emelle some 25
miles from his family's country home. "I'm not sure what Chemical
Waste Management is," he told his wife, "but I think there's a dump
somewhere in Sumter County."

Linda Munoz regarded the dump's presence warily. "Here was
something dangerous that was going to go into my water and into the
air," she said. "I felt violated." In truth, no evidence indicated that
the dump's contents had penetrated seven hundred feet of Selma
chalk and seeped into the water supply, nor did the company incin-
erate wastes that might have sailed across the skies to descend upon

the Munoz home. Nevertheless, Munoz committed herself to closing the dump, and quickly sought the assistance of her neighbors.

When she met Kaye Kiker at a meeting of the Sumter County Historical Society, Munoz asked her to join in gathering information about Chemwaste and its local operation. Kiker was a Sunday school teacher, never before involved in politics. For two weeks, she prayed, asking for direction: "God, if you want me to be a part of this, let me know." "Kaye embarked on the path then," said Charles Munoz. "Irrevocably."

As the sun began to set, we moved to the porch of the Munoz's white wood mansion. Charles disappeared for a few minutes into the kitchen, soon to return with a large tray of drinks. We sat comfortably sipping mint juleps in their iced silver goblets as the pewter sky suddenly blackened; then we listened to the first rumbles of distant thunder that announced the spring storm. Soon, Kaye Kiker arrived, pulling her car into the driveway. Kiker was a very large woman with a soft, melodious voice, and she seemed to shimmer up to the porch. Her white dress billowed in the breeze, and she covered her red hair with an enormous iridescent purple straw hat, its brim wide enough to shelter all of us from the inclement weather. In an instant, the lightning streaked across the sky, rain whipped down upon the porch, and we retreated under the high ceiling of the wood-beamed living room to talk about the early days of ACE.

At the beginning, ACE's membership consisted entirely of white people living many miles from Emelle. But soon Kiker and Munoz realized that the fight against Chemwaste required active support from the black people whose homes surrounded the dump. "Everyone told us we should meet Wendell Paris," recalled Linda Munoz. "But they warned us, too: He's not a very popular man with the whites in this county."

"He's the *enfant terrible* of civil rights here," chimed in Charles. "So he's really much feared, and so on."

Linda bluntly agreed: "The whites hate him."

In fact, when Munoz and Kiker decided to meet with Paris, several white members of ACE immediately quit the group. Even Kiker felt some anxiety.

"I was afraid of him," she admitted. "He was a civil rights person—did he hate whites, or what?"

When the two women undertook the pilgrimage to Paris' office to talk with him about the dump, much of the white community recoiled in shock.

"It ranked up there," Charles Munoz speculated, "with the time Miss Julia, a nineteenth-century humanitarian, arrived in town on the train. A black servant came to pick her up in a wagon and she climbed on and sat right down in the front seat with him, talking like they were good friends or something. It was one major scandal!''

The initial meeting between the white and black environmentalists fared better than anybody could have predicted. Paris had his own suspicions about the dump, and he readily accepted the women's invitation to meet with them in their homes. Eventually, Paris, his brother, George, and George's wife, Alice, became voting members of ACE. Yet no other blacks joined.

"It wasn't like some movie," cautioned Charles, "where all of a sudden the blacks and whites are in the same boat together, merrily rowing down the stream. But it was a watershed for this area—blacks and whites talking to each other."

"What people didn't say about us!" sang out Kiker.

In fact, some people said Kiker and Paris must be having an affair—the only conceiveable reason for their new alliance. Others delivered a stronger message. One evening Kiker's husband picked up the telephone to hear an unfamiliar voice hiss: "We know what your wife's doing. . . ." And then the phone clicked off. "Well-meaning people told my mother," recalled Kiker, " 'If Kaye doesn't leave the blacks alone, she's bound to be murdered.' "

"And I'd never had an enemy in my life!" declared Kiker. "Those calls were the turning point for me. My mama never raised me to be an environmentalist, but this *is* America! I have a right to speak out!"

Booming thunder and a flash of lightning extinguished the electric lights in the Munoz living room. Charles ransacked the cabinets for matches and candles. He finally extracted a long, tapered candle, lit it, and we gathered around its glow.

"The house is haunted, you know," said Linda Munoz. "The lady who lived here was murdered—that's how we got it so cheaply. The thirteenth anniversary of the murder was last week." The wind battered the windows. In another room, the baby cried. Linda left the room and returned moments later with her one-year-old cradled in her arms.

"Some prisoners had escaped," she continued, "and they were known to be in the area. The woman who lived here, alone, was a retired nurse who had worked in a mental hospital. When the prisoners escaped, her sister, who's 94-years-old now, begged her to come stay with her. But she said, 'No, I'm not afraid.' So she went out to church. And when she came home, the prisoners snuck in right behind her. *She just wasn't afraid!*" Munoz said incredulously. Clearly, if the elderly woman had been more careful, more prudently alarmed, more vigilant about the threat, the calamity could have been avoided.

"There will be a terrible, terrible disaster up there," Munoz concluded, skating over her story of the murder 13 years ago to her present fears regarding Chemwaste. "Those pits are already leaking," she asserted, "people will be killed." She stared at the flickering candle.

"That dump simply must be closed down," said Kiker quietly.

A few days before, the two women had been invited, along with John Zippert and the other Federation members, to the meeting we had attended at Chemwaste with Gordan Kenna. We asked about their absence.

"Chemwaste has tried a number of times to get ACE to talk to them," said Kiker. "But we wouldn't do it because they are criminals."

Munoz nodded in agreement. "I have nothing to say to them."

We suggested that ACE's plans to shut down the dump seemed to conflict with the aims of black leaders in Sumter County, who wanted assurances about safety, but still desired the plant open because of its economic benefits.

"Saying that Chemwaste provides jobs and opportunities," maintained Kiker, "is a lie. The tragedy is upon us. We'll never be able to recoup what we've lost because the largest toxic waste dump is in our town." We asked if they had talked with black leaders in Sumter County about ACE's goal of closing down Chemwaste. "We talked to Wendell Paris," said Linda Munoz, "but he had very mixed feelings about it. And now Wendell has moved to Mississippi, and his brother and sister-in-law have gone to Tuskegee." Although ACE had earned a national reputation for pioneering a biracial community organization to fight toxics, it now appeared that the group had not contained a single active black member for over two years.

"We have our hand on the pulse of the situation here," insisted Kiker, "and the blacks are very appreciative of what ACE is doing to oppose the dump."

"The black people," explained Linda Munoz, "were locked into the dump because of economics. When we started this fight, the blacks didn't really know what was going on. It was amazing to me that they weren't up in arms about it. What we found out is that basically the blacks don't know much about the Emelle plant except that their brother-in-law works there and is making pretty decent money." She shook her head in exasperation. "I gave up on this community a long time ago."

"We're a national organization now," explained Kiker. "So there's a different focus."

For some time, the two women talked about their aspirations to influence the world beyond Sumter County. Munoz pointed out that Kiker had collected "an awesome slide show," which she used to

Charles and Linda Munoz, Kaye Kiker, and the nanny, Kate Davis, who helps with the Munoz' children.

inspire people as she traveled around the country. Kiker had been profiled in *Southern Magazine* and *McCall's,* and she had appeared on television's "48 Hours." She had even received the 1988 Presidential Volunteer Action Award from Ronald Reagan.

"This is the greatest thing that has happened to us," insisted Linda Munoz. "To become involved in an issue. We got out there and marched and sang, 'We Shall Overcome,' and held hands. And I realized, Wendell's not a devil. He's a nice person."

"This has opened our eyes," agreed Kiker, "about civil rights."

The next day, we talked to John Zippert about what had seemed to us to be a breakdown in communication and cooperation between Sumter County's white and black environmental activists.

"The people in ACE are well-intentioned people who pride themselves on their purity on the issue," concluded Zippert. "But if they want to influence Sumter County today, they have to be willing to sit down with and take direction from black people. I'm not sure that Kaye Kiker and her group are honestly ready to do that. And if they're not willing to do it, they're not going to have any popular backing to change anything in Sumter County."

-    -   -

In October 1991, the Commission for Racial Justice, directed by Reverend Ben Chavis, sponsored in Washington, D.C., the First National People of Color Environmental Leadership Summit. The central issue animating the five-day convocation was the problem of evironmental racism. By the conference's end, more than six hundred participants, largely African-Americans, Hispanics, and Native Americans, had reaffirmed their commitment to building a national movement focusing on the relationship between environmental concerns, including hazardous waste, and other social justice issues, such as poverty, housing, education, and employment.

"Environmental justice demands that public policy be based on mutual respect and justice for all peoples," proclaimed the 17-point platform ratified by the conference's participants, "free from any form of discrimination or bias." The statement stressed the necessity of minority communities participating as "equal partners" at every level of planning, decision making, enforcement, and evaluation in the siting of toxic waste facilities. Moreover, the conference focused attention on the environment as a civil rights issue that reflected the larger inequities of race in the society.

"The enemy is not Chemical Waste Management in Emelle," concluded Chavis. "The problem is with all the industries that produce these wastes that have to be sent to places like Emelle. The Emelle facility is only a link in a very long chain that begins far away from Alabama. We've learned a lot from Emelle, but it's not just a local battle."

Yet Emelle revealed the fundamental inequities inherent in siting toxic waste facilities in *any* community. Finally, nobody ever volunteered to assume the risks of the nation's chemical overload. Gordan Kenna's plea—"Who do you trust?"—mattered less than another question: Who decides? In Emelle, a 79 percent black town, the critical decisions were still being made by a handful of white people. The volatility of this situation could be measured by the distance between black and white lives: the chasm that lay between Drayton Pruitt, entrepreneur, and Wendell Paris, protestor; between the white activists of ACE who wanted to shut down the dump at any cost, and the black people in Sumter County collected under the banner of the Federation of Southern Cooperatives, who would bear all costs, financial and human, whether the plant stayed open or eventually closed.

And yet, as most participants in the controversy admitted, the battle over Emelle's dump *had* drawn together blacks and whites. Their alliance proved temporary and imperfect; but it was also unprecedented. It had expanded the boundaries of the possible, indicating some hope for future collaboration. It pointed towards the necessity of a national conversation that would draw together the debate over environmental practices and the larger issues of social justice. Although its words might barely be audible, this conversation had already begun in Emelle. In fact, one year later, Kaye Kiker and the Federation's Cleo Askew had linked up, monitoring Chemwaste's actions. "ACE had been out on their own, working alone for quite some time," said Askew. "But then we realized—they've got their constituency, and we've got ours. We're working now to get them all together."

Wendell Paris agreed. "We've really got to look at this thing long-range," he told us one morning in a small cafe outside of town. "We're still fighting the same power structure that's been in place for years. But it's changing."

And then over breakfast, Paris drifted into a long, winding reminiscence about Alabama in the 1960s, during the headiest, most dangerous days of the civil rights movement. He talked about nearby Lowndes County, where Stokely Carmichael had worked in the voter registration drives—and the murder, just over the border in Philadelphia, Mississippi, of James Chaney, a young black civil rights worker

Chemwaste worker wears full body and breathing protection to sample chemicals before they are buried in the ground at Emelle.

from the South, and Andrew Goodman and Michael Schwerner, two white Northern volunteers.

"They didn't know that when the car pulled up behind them with their bright lights on, it was the Klan," said Paris, gulping down another cup of coffee in agitated recollection. "So those boys followed the law and they drove 50 miles per hour, when they needed to push it to the floor and get out of there. What I'm saying is that by the time the movement came around to Alabama, we knew all this stuff. We had developed the know-how to protect ourselves. Blacks started getting elected in Alabama after blacks were killed. Our folks react to clear and present danger.

"More and more," concluded Paris, "the poison in our communities is being seen as a clear and present danger."

We paid our bill, shook hands with Wendell Paris, and together we walked over to our rented car. Paris still had something to tell us.

"It wasn't the civil rights movement that created the images that everybody remembers of Birmingham, Alabama," he insisted, poking his head through the car window as the engine turned. "You got to remember: The problems were there already. It was the foolishness of the Klan's bombing the church that brought Birmingham to an explosion. And the way this country keeps dumping their hazardous wastes in our communities, who's to say that places like Emelle won't explode too?"

"Tonight Feels Something Like the Fall of Saigon"
—McFarland Today

Carlos Sanchez with his family. Carlos was not included in the statistics and studies of the McFarland childhood cancer cluster, since he was one year over the age limit when his tumor was discovered.

On October 24, 1991, the state-appointed Scientific Advisory Panel met at Veteran's Hall in Bakersfield, about 50 miles from McFarland, to announce its final conclusions regarding the town's mysterious childhood cancer cluster.

"Today we are going to put the entire thing into perspective," announced Dr. Lynn Goldman, the California Department of Health Services epidemiologist who headed the investigation and now chaired the meeting. "We're going to give the big picture. And at the end of the day, we'll have a formulation about where we should go from here."

As the committee's nine members took their seats around the long rectangular conference table at the front of the room, a stream of observers slowly filed in and filled up a dozen rows of metal folding chairs. The audience consisted chiefly of media, since most of the families living in the farming communities studied by the Department of Health Services preferred to wait for the less technical presentation scheduled for later that evening in McFarland.

Reporters from the *Los Angeles Times*, the *Bakersfield Californian*, and the *Record* in Delano greeted one another, nodding perfunctorily or quickly shaking hands. They had been following this story for seven years now. A young television news reporter with his perfect black coiffure and flashy blue suit breezed back and forth across the back of the room while his camera crew labored with their equipment. Occasionally, the television reporter zipped out the door to scan the scene for pictures in front of Veteran's Hall.

Outside, a contingent of 50 United Farm Workers supporters formed a desultory procession, bearing the union's black and red banners and waving bilingual protest signs—No Grapes, *Uvas No*. Mostly middle-aged and elderly—lifelong laborers in California's fruit and vegetable fields—the marchers sported sun-worn straw hats, clean flannel shirts, heavy denim work pants and boots. The first UFW supporter to arrive on the scene hours earlier had been 64-year-old Genaro Zavala. He thought the meeting involved contract negotiations with *los rancheros*, the growers. A huge Mexican sombrero wobbled atop Zavala's head, giving the impression that his pole-skinny, sun-crisped neck could barely support the hat's tremendous weight. Across the sombrero's brim, Zavala had scrawled in thick black letters: "*viva la causa*" and "*Zapata*."

Most of the protestors spoke little English, and hardly any of the reporters clustered around them spoke Spanish. The UFW procession was composed entirely of Mexican-American supporters waiting patiently for the arrival of Cesar Chavez. Over the course of a half hour, the ranks swelled to one hundred marchers. New signs and banners appeared: *Boycott Pesticides* and *Pesticides Cause Cancer*. A line of skulls underscored the signs' inscriptions.

"Why are you here?" asked one of the local newspaper reporters.

"Mr. Chavez will tell you later," replied a demonstrator. "He's coming soon and he will tell you why we're here and everything you want to know."

In Spanish, the demonstrators' spirited conversation revealed that they knew exactly why they were there. The demonstration's lead

man dropped his megaphone to his side and launched into a booming speech full of passion and solidarity, denouncing pesticides as the
cause of the children's cancers not only in McFarland, but also Earlimart—the cancer cluster recently unearthed by the UFW and subsequently confirmed by the state. Under the shade of a tree on the
lawn outside Veteran's Hall, the protestors sprawled, listening attentively and cheering the familiar harangue: *la causa* of the 1990s.

Back inside the hall, Dr. Lynn Goldman called to order the Scientific Advisory Panel. The group included epidemiologists, researchers, and pesticide experts from the federal Centers for Disease Control, the California Department of Health Services, the University of
California at Berkeley, UCLA, and the Kern County Health Department. The conference tables were arranged to form a horseshoe at
the front of the room, opening up to the audience. The panel's further efforts to appear approachable and relaxed seemed somewhat
studied and forced—the male scientists casually attired in sports coats
and slacks, the women foregoing the androgenous power suits of the
urban business world and academe for long, bustling dresses or skirts,
and peasant blouses presumed more suitable for a community meeting in the Central Valley. The Department of Health Services' community liaisons had emphasized the importance of shaking the image
of professional detachment that irritated and confused many people
from the cancer cluster towns. Community people tended to interpret
the professional virtues of scientific skepticism as aloofness or chilly
disregard for their plight.

The panel began the meeting with an extensive presentation of
its findings. They had looked for other towns in the valley's four counties that might also have unusually high rates of childhood cancer.
Overhead projectors splashed across the front wall the magnified illu-
minations of plastic overlays diagramming the scientific research and
statistical methods. Following a good deal of technical discussion,
largely unintelligible to most of the audience, the panel moved on to
its central conclusion. According to the study, no community located
within the four counties that composed the state's agricultural heartland demonstrated elevated rates of childhood cancer, except for the
three towns everybody already knew about: McFarland, Fowler, and
Rosamond. Earlimart, the Hispanic farming community near
McFarland that had previously been identified by the state health
department as a cancer cluster, did not qualify in the four-county
study. One of the town's cancer cases had fallen outside of the study's
designated time period, and another had occurred beyond the town's
official census tract.

"The three communities with a higher incidence of cancer were the ones originally identified by the communities themselves," observed Dr. Goldman, casting a wry smile to the audience. "For an epidemiologist, that's truly humbling. Our epidemiological methods weren't any better than the communities." Finally, the four-county study was the last link in a series of fruitless investigations that had cost hundreds of thousands of dollars during the course of five years. The health studies had raised the hopes of worried parents, dashed them, and then repeatedly raised them again—only to finally have found nothing. Outside, the UFW demonstrators encircled the building. The sound of their stamping feet and rambling chant rose sharply. "Boycott, boycott!"

A staffer from the Department of Health Services shut the door. The noise diminished slightly.

"*Abajo los mentirosos!*" sang out the marchers. *Down with the liars.*

"Such *charming* people," said Merry Ellen Alls, the McFarland science teacher, who had taken a seat in the audience. Her eyes stared straight ahead at the conference table, but her mouth formed an expression of skepticism and impatience. She raised her hand, and rose to address the committee.

United Farm Worker members protest outside of the health department meeting in Bakersfield.

"Speaking for the people of McFarland," she said, "and I guess it's quite rare that I would ever think of myself as speaking for the people of McFarland, I would like to thank you all for the time and effort you've put into this work. And I want to let you know—understanding that you've heard very little like this over the years—that we *appreciate* all you have done."

"*Boycott la junta!*" resounded the muffled chant from outside. *Boycott the meeting!*

Goldman smiled politely at Alls, and then asked for other comments from the audience.

A slender, nut-brown, sun-baked man in his late fifties—from all appearances, the only Mexican-American in the room—rose from his chair and asked about the safety of McFarland's drinking water. He observed that a well had been closed because of contamination from the pesticide DBCP. Why didn't the committee close all the other wells? "Are you waiting to see how many people will get sick?" he demanded.

Goldman explained that the water company had closed the well, but she agreed with the man that the action had taken far too long. She empathized with his frustration. She decried the slow pace of government. And she admitted, skillfully, without assuming personal or collective responsibility, that the worlds of science and government still had much to learn about working together effectively.

The remaining questions and comments from the audience proceeded as expected. The panel members listened attentively, bolted straight up around the conference table. None of the scientists looked rankled or surprised by the note of skepticism that rose like a tremelo in the voice of almost every speaker. Whatever the criticism, they had heard it all before.

Then a 21-year-old man, with a round, smooth face that made him appear much younger, stood up in the back row. He wore an uncomfortably fitting suit, his brilliantine blonde hair slicked back. He looked and sounded like an exceedingly polite and respectful young man who took very seriously everything he had heard from the panel. Over one eye, he wore a huge black patch.

"My name is James Sherrill," he announced with the clarity and courage of much rehearsal, "and I live in Rosamond, and I had cancer." '

And then James Sherrill broke down into sobs. "There's people *dying* out there. . . ." he cried. "That's why I'm here today. I had cancer. I got cancer in my eye and my nose. My friends died of cancer. My buddies in school, they're dead now." Sherrill halted, wiped his face, collecting himself; clearly, this is not what he had intended. "I appreciate that you guys tried," he told the panel. "But *something's* got to be done."

For the first time, the scientists looked unnerved. People didn't usually stand up at meetings of the Scientific Advisory Panel and break into tears. In fact, this meeting marked only the second time that the committee had opened its doors to the general public. The members of the Scientific Advisory Panel gazed uncomfortably across the room at Sherrill. What could they say to him? They no more knew what had caused the cancers in Rosamond than they did in McFarland. For a long moment, nobody said anything.

"BOYCOTT! BOYCOTT! BOYCOTT! BOYCOTT!"

A UFW protestor placed his battery-powered megaphone up against the outside of the door, shouting the slogan at top volume. His scratchy, amplified voice penetrated the thin membrane of the cool, cream-colored walls that had sealed off the meeting room from the world outside. Suddenly, the chanting of the protestors and the sobbing of the young man with cancer drew together the parallel worlds of scientific inquiry and public outrage. The experts sat at their horseshoe of rectangular tables with microphones set in front of each of them. They stared at the walls, slightly irritated and thoroughly confused.

Then Steve Schilling, the director of the health consortium that ran the new McFarland medical clinic, stood up from his seat in the audience. His deep voice boomed far louder than the microphoned protestor. "I want to know why the Department of Health Services, in their great wisdom, has *cut twenty-five thousand dollars from our operating expenses?*"

The question was a statement, and nobody could really respond adequately. Members of the Scientific Advisory Panel expressed their surprise and sympathy. They agreed that McFarland needed the clinic. Most rural California farming communities suffered appalling gaps in health care. The problem, as everybody in the room understood, was money.

Then the director of the California Tumor Registry stood up from the audience and declared that her office budget had been cut by 30 percent—after operating for only a few years. "At this point, each of us is doing the work of two people," she asserted. "And you know that your studies will be garbage without the cancer registry."

Again, the panel members nodded their heads in agreement. Dr. Lynn Goldman revealed to the audience that the annual budget of the entire Department of Health Services had been cut by 20 percent, affecting every important health program in the state of California. This talk of the state's lack of funds segued perfectly into the panel's discussion as to whether they should recommend further studies of McFarland and the other cancer cluster towns.

"I understand that the committee wants to find an environmental cause," spoke up Dr. Robert Spear from the University of California at Berkeley. "But how much more are we going to spend? Over the years, there's been a tremendous effort to find a source of exposure causing these illnesses. It hasn't been found. I have no faith that additional investment of resources, regardless of the amount, could tell us anything we haven't seen."

Spear's opinion seemed widely shared. The entire panel felt extremely reluctant, particularly in light of other pressing health needs—exemplified minutes before by testimony from the McFarland clinic and the state tumor registry—to ask the State of California to spend more money on another expensive study in McFarland that would almost certainly produce no conclusions, no culprits, and no assurance that the lives of future children would be protected.

Outside, the sound of the protestors swelled up and crashed through the barrier of stucco walls like a wave: "*Boycott, boycott!*"

Dr. B. A. Jinadu, of the Kern County Health Department, pointed out that the previous studies had not nearly measured the full range of possible toxic exposures. "We haven't put one pesticide into the test tube, and then added another, and then taken in the human factor," he insisted. The members of the Scientific Advisory Panel shifted uneasily in their seats. "This may not be as easy as I suggest," Jinadu admitted.

"We're not even *close* to estimating the effects of exposures to combinations of chemicals," insisted Dr. Robert Haile of UCLA, "which is how people are really exposed." Then the expression on his face radiated alarm: "I'd hate to think that anybody is getting the impression that we're giving the environment in McFarland a clean bill of health!"

"*Boycott, boycott!*"

An elderly woman from McFarland bustled into the room, apologized to nobody in particular for being late, and took an empty seat in the audience. When she asked the person sitting next to her what the panel was talking about, her neighbor rudely shushed her into silence. The woman popped up from her seat and declared the panel's report to be "rubbish." Both the panel and the state health officials had heard from this woman many times in the past; the encounters always proved extremely time-consuming, and always led nowhere. This morning, the woman's critique similarly wove through an extensive labyrinth of illogic, full of oblique references to other studies. Nobody in the room could follow the argument. Finally, Gold-man cut her off. "*Thank you,*" she said.

Another McFarland resident stood and asked, simply enough, if McFarland, or any Central Valley town, was safe to live in.

"Scientifically, that question is impossible to answer," responded Dr. Haile. The room stirred slightly.

Dr. Jinadu, the Kern County health officer, appeared shaken by his colleague's response. "If it's not safe," he declared, "we should close down the town."

"No," interrupted Haile, "we can't say it's safe to live in any town anywhere. The problem is the way the question is phrased." Haile went on to explain that scientists could not guarantee the absence of danger in McFarland or anyplace else. They could simply say that they had not found anything in the environment that presently indicated danger. But this was not the same thing as assuring its safety.

For a moment, the room fell into an uneasy quiet as the audience strained to grasp the full implications of Haile's speech. The scientists, it appeared, would not—could not—offer the kind of unequivocal answers that most people longed to hear.

Dr. Spear broke the silence with his own reaction to the perils and frustrations of scientific uncertainty.

"Over time we're just going to have to get rid of pesticides," said Spear, "whether through integrated pest management or whatever. But we're never going to have the scientific evidence."

Spear's comment lured the discussion away from the realm of theory, and back toward practical matters. The panel spoke among themselves for some time about McFarland's clinic, which did not require scientific certainty to prove its importance to the community. Given the failure of all the past health studies, the panel then dis-cussed redirecting funds to support the clinic's ongoing work.

"Do you have the authority to do that?" Merry Ellen Alls asked from the audience.

The sound of the protestors rose once again: "*Boycott, Boycott!*"

"We can recommend what we want," replied Goldman. "We don't have control over a pot of money, but we can make recommen-dations."

Goldman pressed the panel for an official statement. Its members did not want to commit themselves.

"*Boycott, boycott!*"

"Should I take it from our committee," she pressed on, "that while we agree there are some methodological efforts that we could make, there's nothing more we should recommend regarding further scientific studies of McFarland? Should I take it that that's our recommendation?''

The members of the Scientific Advisory Panel uttered some meager sounds of agreement, barely nodding their heads.

"BOYCOTT, BOYCOTT!"

Moving on to the subject of Earlimart, the committee faced another difficult decision. Although the four-county study had not identified the town as a cancer cluster because two of the cases failed to fit into the study's strict protocols, the California Department of Health Services nevertheless regarded the community's childhood cancers as significant and worthy of attention. The question remained: how much attention, and at what cost?

Finally, Dr. Haile formulated a long paragraph regarding what to do next in Earlimart. It boiled down to recommending against a McFarland-type study in which every possible exposure route would be pursued. Rather, the committee encouraged the "hot pursuit" of the most likely leads related to the cancer's cause. They also supported continuing monitoring for new cases. "No extensive investigation of Earlimart is recommended," summarized Haile, "because it is unlikely that it could solve the puzzle."

Lastly, the committee faced the inevitable decision of whether to dissolve itself and deem its work concluded. Unceremoniously, the committee did just that.

"McFarland remains a mystery," proclaimed one member.

■   ■   ■

In front of Veteran's Hall, the United Farm Workers' president, Cesar Chavez, finally arrived, drawing outside the full media barrage.

At Chavez's side stood Guillermo Robles, the father of Miriam Robles, an Earlimart child with cancer, who would die within the year. Robles made a brief statement about his daughter's illness, followed by his conviction that pesticides were the cause. Shyly, he stepped aside, uncomfortable with the flashing cameras and furious scribbling of the reporters. Chavez wrapped an arm around Robles' shoulder, comforting him in his brave effort; then, the United Farm Workers president stepped forward to the microphones.

"They're playing politics," proclaimed Chavez, speaking rapidly in English about the Scientific Advisory Panel still assembled inside. "It's like a Watergate." Chavez reminded reporters that the earlier McFarland studies had turned up at least one important finding: The children with cancer in town were shown more likely than the kids without cancer to have parents working in the fields, where pesticide exposures proved routine. A review of the survey had revealed a less striking statistical correlation, although it still pointed to a possible lead. A more thorough study of the relationship between childhood cancer and parental occupations in the Central Valley was thought to cost at least another million dollars.

Chavez also detailed the results of a recent study conducted in Los Angeles in which 123 children with leukemia were compared with an equal number of healthy children. This case-control study, said Chavez, showed that kids whose parents used pesticides in their homes and gardens were many times more likely to develop cancer. "Why can they find this out by case-control studies in an urban setting, but they do not do the same in a rural agricultural setting?" he demanded. "If the state did the kind of study we suggest, they would find out what we know—that there is a direct relationship between cancer and pesticides."

Chavez's insistence on the importance of a case-control study seemed strangely removed from the events of the day. The seasoned organizer was promoting a technical solution to resolve the essentially political problem of pesticide use—despite ample evidence that studies of the sort he championed seldom inspired the profound kind of changes that the union sought in California agriculture. Moreover, as Chavez must have known, the Department of Health Services had conducted a case-control study in McFarland years before, at the very beginning of the investigation. More to the point was the official statement issued by the union following the meeting.

"Farm workers know it isn't an accident," read their press release, regarding the inconclusive health studies. "It isn't a statistical quirk. It's the pesticides. And until farm workers see results of scientific studies showing that exposure to cancer causing pesticides is *not* a factor in childhood cancer in agricultural towns, the state should *not* allow their children to be used as guinea pigs."

After Chavez concluded his speech in English, he begged off the reporters' questions until he ran through everything once again in Spanish, for the benefit of the union supporters clustered around him. In the meantime, some of the photographers drifted off to the fringe of the crowd, rediscovering Genaro Zavala in his Zapata sombrero. Whenever a camera crew assembled at his side, Zavala slowly, almost deliciously, rolled up his pants leg to the knee, revealing a ghastly skin rash. Through a translator, Zavala explained that he had worked as a picker and pesticide applicator—jobs affording the most dangerous chemical exposures. "I used those chemicals all my life," he said, "and sometimes the bosses even forced me to use them at twice the strength I knew they were supposed to be mixed at. The boss would tell me there was no proof—*no es cierto*—that the chemicals were dangerous. But to me, *es cierto*."

Out on the front lawn, under the ponderous beat of the sun, James Sherrill, the young man from Rosamond with cancer who had addressed the panel inside, also spoke on-camera to the world. "I came to this meeting because I am very concerned about the environment," he declared, his emotional testimony inside the auditorium now ironed flat into a platitude for general consumption. "I want to be sure they clean up the environment so it will be better for the children of the future."

After the interview with Sherrill, Karl Schweitzer, the television news reporter for KERO, Bakersfield, assumed a solemn expression and told his listeners: "I've been covering this issue for some years now and the only thing that is clear to me from today's meeting is that it all still sounds the same. The health department has been looking for a needle in a haystack, and the needle has disappeared."

— — —

Cesar Chavez addresses the press on the lawn outside the health department meeting.

At 7:30 that evening, Dr. Lynn Goldman and other members of the Scientific Advisory Panel gathered at the McFarland Community Center to present their findings and to field questions from local residents. About 25 people showed up for the meeting. Tina Bravo did not attend because she could not bear to hear the officials admit that they could find no cause for the cancer that had killed her son, Mario. Connie Rosales was in the hospital with a broken leg. Borjas Gonzales claimed that after the death of his son, Franky, he had kicked the health department investigators out of his house; he didn't want to hear anything more about cancer, cancer studies, or the California Department of Health Services.

As the meeting began, several members of the committee sat at a table in the front of the room, while the other scientists took seats spread throughout the audience. Dr. Rick Kreutzer, from the state health department, moderated. Kreutzer wore a charcoal coat, purple shirt, and purple tie flaring into green, with his hair tied into a ponytail. Underneath the table sat his shoulder bag, instead of a doctor's briefcase. As Kreutzer spoke, each sentence was gracefully translated into Spanish by a a state interpreter. "I want to take this opportunity to respond to a rumor," announced Kreutzer. "There wasn't a new child cancer in the last six weeks. It was an adult."

This announcement hardly qualified as good news, and nobody in the audience appeared particularly relieved. Peggy Reynolds, a Department of Health Services epidemiologist, set up the overhead projector and illuminated the overlays as she outlined the results of the four-county study. She reiterated the committee's conclusion that only McFarland, Fowler, and Rosamond evidenced high rates of cancer, but no cause could be found. The crowd showed no reaction.

Kreutzer explained why the Scientific Advisory Panel recommended against pursuing a McFarland-style study in Earlimart. He assured the audience that the Department of Health Services would continue to follow up leads.

Again: no visible reaction. Then Kreutzer called for questions.

A huge, hulking man, standing six-feet-seven-inches, rose from his folding chair and identified himself as Gilbert Flores, from Delano. He asked why two teenagers and a child with cancer in his town weren't included in the four-county study.

Kreuzter explained that these recent cases, occurring in 1991, fell outside of the time period investigated by the state.

Another man stood up. He was slender, in his thirties, and he spoke only Spanish. "I understand perfectly well that this is a study of childhood cancer," he told the panel by way of the translator. "But what's going to happen to the adults suffering with cancer? My wife has cancer. That's why I am asking. What is going to be done?''

Kreutzer responded gravely, sympathizing with the man. There seemed little that the doctor could offer by way of practical advice to the stricken woman or her husband, but nevertheless Kreutzer talked for several minutes to the rest of the audience about the importance of early cancer detection among adults, and the necessity of seeing a doctor. Then he ran through the early warning signs of cancer.

At the back of the room, Martha Salinas raised her hand. She wore a plain white skirt brushing over the floor, a purple blouse, and a white sweater with flowered embroidery. With her deep-set brown eyes, radiant smile, and her black hair streaked with grey lilting upon her shoulders, she looked striking and impossible to ignore.

"Rick," said Salinas, addressing Kreutzer by his first name, "I think it's important to mention the other signs of cancer." And then she did just that, enumerating the symptoms of nausea, dizziness, constant vomiting, fevers, rashes. . . .

Kreutzer finally cut her off, agreeing that people with these symptoms should see a doctor.

A health department community liaison waved above her head the handout listing the symptoms of cancer. A stack of them sat upon a table in the back of the room, along with other printed information sheets about the health studies.

Martha Salinas rose once again to upbraid the panel for not involving the federal Centers for Disease Control. This time, Lynn Goldman pointed out that Dr. Matthew Zack of the federal Centers for Disease Control, who was now sitting in the room, had been a member of the Scientific Advisory Panel. Salinas then upbraided Zack and the rest of the panel for ignoring the "pain of residents, the new victims on the block."

"You can go home and study the reports all you want," she told them, "but the victims still stay here."

Asa Bradman, a scientist with the Department of Health Services, tried to get the meeting back on a more focused direction. He moved to the center of the room where he presented the third and final stage of the McFarland health study.

As Bradman spoke, Martha Salinas frowned and wrinkled her forehead, alternately drumming her fingers on a legal-sized yellow notepad and scribbling comments in its margins. Then she raised a hand and spoke up: "Asa, I think what's most important is. . . ." Salinas launched into complicated speculation about the inadequacies of the soil studies. Bradman insisted that the soil samples had been drawn at various levels, measuring at depths of one inch, one foot, and three feet—and they produced no plausible suspects of exposure.

"You say that the water is not dangerous," countered Salinas. "But I have information from the EPA that nitrates cause cancer."

"EPA doesn't classify nitrates as carcinogens," contradicted Dr. Lynn Goldman, growing alarmed about the direction the question period had taken. She briefly explained the relationship between nitrates and nitrosamines. The audience grew restive.

"It doesn't happen overnight," asserted Salinas. "I have those EPA reports."

In a soft, mild voice, Department of Health Services scientist Sharon Seidel offered to review her reports.

"The water isn't clean," insisted Salinas, ignoring her. "It has arsenic, mercury, and DBCP."

"There was arsenic contamination in well number five," admitted Bradman.

"And six," said Salinas.

"The DBCP was found in well number five."

"And oil."

Bradman agreed that the wells had been taken out of use by the water board when contamination was discovered several years earlier.

"What I'm saying," summarized Salinas, persuaded that the panel had not yet grasped her point, "is there's a lot of work that you've yet to do."

Kreutzer tried to stop her. "Anybody who can provide us with additional environmental information, please contact us. Thank you."

Salinas' fervor inspired several other questions from the audience. Kreutzer tried his best to answer them, though nobody really felt satisfied. But the audience sensed the sincerity of his effort, and the room calmed down. "I think we've done the best job possible," Kreutzer admitted finally. "If you don't like it and have information, tell us."

Goldman stood, attempting to wind up the evening.

"We'd like very much to hear your comments to the study over the next two months during the review process," she told the audience. "We value your comments just as we value the comments of the Scientific Advisory Panel. We didn't find the source of the cancers. But that doesn't mean we're not concerned about the pesticides, nitrates, and other problems that you've raised tonight. Many people have said we don't want to find the cause of the cancers. My belief is that we have done everything possible with the tools of science. We're disappointed."

In a final effort, Martha Salinas rose once again from her chair to demand that the panel explain its statistical methods to the audience. The panel members looked astounded that she would be asking at this late moment in the evening for a short course in statistics. Then Salinas insisted that only poor people were counted in the studies, after being identified from welfare rolls. Goldman flatly denied the charge. Her expression seemed to read: *What next?*

"McFarland has a dark cloud over it," Salinas rolled on. She suggested that doctors refused to treat McFarland kids because of the stigma attached to the town. "Thank you," Kreutzer told Salinas, with an exhausted note of finality. The strain of this last effort had grown palpable; the committee seemed to be dissolving before the community's eyes, the sickly limelight it cast over McFarland now fading. "Thank you," he repeated.

As the audience began to file out of the auditorium, Martha Salinas stood alone in front of a local television news camera and explained to the world that McFarland should be regarded as another Times Beach, Missouri, or Love Canal. On camera, Salinas appeared avid and alert, though perhaps slightly too passionate to be entirely convincing.

As Lynn Goldman walked out the door of the McFarland Community Center and into the parking lot to find her car and drive home to the San Francisco Bay Area, she spoke wearily to her staff. "I fear that to the community, tonight feels something like the fall of Saigon," she said. "The helicopters have swept in and picked the U.S. government people off the rooftops—and we've left them behind in the dust to fend for themselves."

-　　-　　-

The next day, the *Bakersfield Californian* ran the following headline: "STATE ABANDONS STUDY OF MCFARLAND CLUSTER." The story also informed readers: "Likewise, the state will not conduct an extensive search for the cause of a six-child cluster in Earlimart. . . ."

At her office, Lynn Goldman raged over the story, protesting to her staff that no matter how correctly the Department of Health Services conducted itself, most people would deduce from the newspapers that the state's behavior had been casual and cavalier. Fortunately for the department, the newspaper accounts failed to mention one final failure of the five-year-long pursuit.

Earlier in the year, the blood samples taken from McFarland's children during their health screening exams—and stored for possible future use in detecting the cause of the cancer cluster—had perished.

"The refrigerator broke down and the blood thawed out to a nice comfortable seventy or eighty degrees," explained Dr. Kreutzer. "The alarm didn't trigger for days. So we don't have those specimens to use. They're gone."

- - -

In the spring of 1992, a new play about a mysterious childhood cancer cluster plaguing an Hispanic, Central Valley farming town opened at the Mission Cultural Center in San Francisco. For two months, *Heroes and Saints*, written by Cherrie Moraga, drew capacity crowds and earned rave reviews from local theater critics.

In Moraga's poetic, powerfully performed play, McFarland was rechristened McLaughlin. Several of the production's main characters were drawn from the people who played leading roles in McFarland's own tragedy.

The central character, Cerezita, was a teenaged girl, born without limbs, who glided around stage on an electric wheelchair—an older, feminized version of Felipe Franco, the boy from Delano whose mother blamed his birth defects on her exposure to pesticides in the fields. Doña Amparo, the activist neighbor who rallied residents to confront the unbridled use of pesticides in the fields where most people in town worked, was acknowledged by the playwright to be a "tribute" to both Dolores Huerta, vice president of the United Farm Workers, and Martha Salinas, "a mother and one of the chief organizers in McFarland who opened her home and the homes of other McFarland families to me."

The first scene of *Heroes and Saints* showed the body of a dead child being draped upon a cross in the grape fields by a band of children wearing masks fashioned like death skulls. At no time did the play broach any doubts that the children with cancer had been made ill by pesticides. From the opening of the play, the dimly lit vision of grape vines rose up in the gloom like the barbed-wire trenches of a World War I battlefield. Relieved of the ambiguity inherent in McFarland's real life tragedy, the artistic recreation of the town bore down with utter certainty upon the source and moral culpability of the children's deaths. In art—at least in *Heroes and Saints*—there seemed less room for uncertainty than in life.

"Secrets can kill sometimes," explained one character.

In the last act, the full cast of characters marched out into the grape fields to noisily protest the pesticides sprayed overhead by hel-

Mayra Sanchez and her parents.

icopters. One mother screamed to the skies that the helicopters should just swoop down and kill them now. "It would be faster!" she cried. In the final moments of the play, the helicopters did just that. The nightmare of McLaughlin bled into the headlines of Guatemala and El Salvador: The growers were transformed into death squads, raining down upon the innocent protestors volleys of machine-gun fire instead of aerated poisons.

Although it was a powerful moment, the exaggeration also felt unpersuasive. Only the most surrealistically poetic license could equate the growers of California's Central Valley with Salvadoran death squads. The nightmare vision undercut the waking tragedy.

In truth, the play's emotional climax took place somewhat earlier in the last act—during a scene portraying a demonstration on the steps of the capitol in Sacramento. In an extremely heartfelt and disturbing moment—one of those rare times in the theater when the entire audience seems drawn together for an instant in their collective anguish and dread—a group of children from McLaughlin stepped forward, one-by-one, holding the photos of the children who actually got sick or died in McFarland and Delano. On stage, the young actors read some of their names. . . .

Tresa Buentello	Juan Gonzales
Franky Gonzales	Tracey Ramirez
Adrian Esparza	Sandy Perez
Carlos Sanchez	Kiley Price
Johnny Rodriquez	Mayra Sanchez
Felipe Franco	Mario Bravo
Angela Ramirez	

Heroes and Saints was held over for two weeks, and closed in May 1992. Although the producers discussed bringing the show to McFarland, the complexities of moving the sets, arranging for equity waivers from the actors' union, and securing a sufficiently large theater to mount the production finally stalled the plans.

McFarland would have to live with its own story.

-14-

The Virtues of Messy Democracy and Uncertain Science

Nova, Ohio.

Despite McFarland's tragedy and apparent failures on the local front, its message to the country at large remained profound.

True, the scientists and health officials did not find the source of the children's cancers. Nor did the community form a lasting, powerful political organization armed with a coherent agenda for action. Nor, in the end, did *anybody* deeply involved with the long ordeal—from the state health department's staff to the afflicted families—feel satisfied with the outcome.

But to dwell upon McFarland's failings is to obscure its greater significance.

Most important, McFarland stood as an *example*: a working model of the fear, confusion, pain, and uncertainty likely to affect even more citizens as the boundaries of toxic America expand.

For three years, we had traveled back and forth across the country, visiting scores of small towns and rural communities. The fiercest fights over toxic contamination were being waged in the places where the American compact proved most fragile. The nation's ingrained inequities of race, class, and gender were being emphatically accented by battles over chemical wastes in minority communities, such as Emelle, Alabama, and in working-class and farming towns, such as Yukon, Pennsylvania, and Nova, Ohio. Almost always, the most effective warriors turned out to be women.

Yet, for all the hardheaded pragmatism and shrewd theatricality of the grassroots anti-toxics movement, its crusaders remained stubborn idealists. Their toxic rebellion had spread across the country with great speed, drawing together thousands of ordinary Americans, precisely because of their persistent, fundamental, and extremely durable faith in democracy.

"No one can contemplate American democracy in the late twentieth century without experiencing the dissonance between the idea and the reality," observed political analyst William Greider. Yet "some citizens, the energetic minority who still believe, struggle to resolve these contradictions. . . . Behind the empty shell of formal politics, the nation is alive with democratic energies. People are still pursuing the universal impulse for political self-expression. Disconnected from power, they are still searching to find it."

But rehabilitating American democracy requires "much more than reforming the government," as Greider also shrewdly noted. It means that "citizens at large must also reinvent themselves." The grassroots toxic rebellion has enabled the ordinary, disaffected, disconnected residents of America to first reimagine, and then reinvent themselves—as citizens, reclaiming all the rights as well as the responsibilities of democracy's untidy accommodations. In practically every community that had attempted to speak forcefully and directly to power, we found a complicated, contradictory, and ultimately, exhilarating demonstration of American democracy at its messy best.

The reinvented, newly invigorated citizens of toxic America—so effective in their local battles, though largely ignored in their national significance by politicians and the press—have much to teach the rest of us. With the moral support and practical aid of their friends and neighbors, they have grappled with the uncertainties of science, pressed up hard against a resistant political establishment, and overcome their own lack of confidence and entitlement to speak out for commonsense solutions to complex, often immobilizing problems. They have demanded accountability from the experts and decision makers, and offered as the price of the bargain their own immersion into civic life.

In fact, it is these ordinary citizens' plain-spoken case for drastically reducing our nation's toxic overload that remains the most persuasive of arguments. Given the present uncertainties and future risks of toxic exposures, they have chosen to protect themselves against tomorrow's darkest vision by taking action now. And while the prospect of some petrochemical apocalypse remains unverifiable, the evidence of mounting risks continues to accumulate. The battles waged by citizens in toxic towns throughout the country may only presage a much larger war to come—if we ignore the message broadcast from places such as McFarland, California; Yellow Creek, Kentucky; Woburn, Massachusetts; and countless other communities.

Yet most industry and government representatives, and many public health officials, argue that exposure to hazardous wastes has not and will not create a genuine public health catastrophe. While acknowledging *some* substantial health effects, the official position states clearly that towns like McFarland should be regarded as inexplicable exceptions. National policy, they reason, cannot be based on fear engendered by a handful of mysterious mishaps. And expenditures of scarce federal and state funds might well be better used to vaccinate against meningitis, or perform research on AIDS, than be spent by the billions to clean up deposits of toxic wastes that may pose much less threat to our national health than dozens of other scourges.

Still, as we traveled throughout toxic America, we could not shake one damning conclusion: As the national toxic overload continues to mount—and it *is* growing, despite a marginal decrease in the production of industrial toxics in recent years—there can be no guarantee of safety. The diseases of the future might indeed be chemically induced cancers, birth defects, and neurologic abnormalities exacerbated by the unfortunate patience of a generation that waited for more evidence before taking action. In any case, the regulations, legislation, and government agencies now in place to reduce the outpouring of toxic chemicals and clean up our gargantuan accumulation of hazardous wastes simply are not working.

The citizens of small-town toxic America have recognized that in a democracy the formation of public policy must rely on common sense and intuition, as well as on scientific proof and bureaucratic imperative. The intuition of toxic America is that steps must be taken immediately to stem the toxic tide.

So far, the grassroots toxics movement has addressed the toxics problem most successfully by utilizing its negative power. The nation's toxic avengers have blocked the construction of waste dumps and incinerators; they have thrown countless uncooperative politicians out of office. The question remains as to whether the objections of rural, working-class America will lead to a louder, longer, more conclusive national reckoning over the future of hazardous chemicals—a reckoning that will also touch the locus of conventional political power in America's increasingly contaminated urban centers.

Today's toxic crusaders have shown their determination to forge the necessary alliance between common sense, rigorous science, and democratic enthusiasm that can produce a wiser, more informed national policy on the use and disposal of dangerous chemicals. Community activists have begun to show the commitment and flexibility required for a collaboration between their own ranks and the best scientific minds; they are struggling to work with the experts without abandoning their fate to them.

Science may someday substantially improve our ability to safely handle hazardous chemicals. For the present, both the understanding of their effects on human health, and our ability to judiciously dispose of them, remain limited. No doubt this gap between aspiration and accomplishment means that much of the energy recharging the contemporary environmental movement will continue to be generated from the bottom layers of American society. Without question, the grassroots movement drives the toxics issue—not the large, mainstream environmental organizations, not the scientists, and certainly not the government; the influence of the grassroots should not be underestimated.

Because finally, the lessons of America's small-town toxic rebellion extend far beyond the important, but limited issues of hazardous chemicals, public health, and the environment.

In truth, the deepest, most intractable problem confronted by America's small-town toxic rebels has been the pervasive disengagement from civic responsibility that has come to characterize American life in recent decades.

At its best, the toxic movement demands that ordinary Americans re-engage; that we take responsibility, both individually and collectively, for our personal and societal choices. These ideals have not fared well over the last two decades; indeed, they have never approached their full potential in all of our nation's history. Perhaps small-town toxic America can help resurrect, nourish, and sustain this core of democratic necessity.

While this task will inevitably prove difficult, full of squabble and strife, it also remains our best hope for the future—not only in terms of resolving controversies involving hazardous chemicals, but also in addressing the entire range of problems now facing the nation.

Until then, the fear of and fight against toxic wastes will continue. The issue will remain both metaphor and palpable reality, a way of perceiving the nation at its worst, while remediating its problems through the best of democratic faith and action.

- - -

Throughout our travels in toxic America, we met many people whom we came to respect and admire. As a group, they proved to us that the heartland virtues of righteous indignation, stoicism, ingenuity, vigor, empathy, and hard-rock determination were still abundantly alive throughout the nation.

But there was one encounter with two atypical toxic crusaders that kept coming back to us in memory over the course of writing this book.

The scene that kept haunting us occurred on our last day in Yukon, Pennsylvania—where we had been talking with community activist Diana Steck and members of her local anti-toxics organization, CRY.

Early in the morning, Diana had led us up into the woods above her town's graveyard so that we could view the local toxic waste dump that CRY had been fighting. If we could just see the enormous pool of toxic chemicals, she insisted, we would understand what her community was up against. The *idea* would assume substance.

Diana led us directly up the hillside, into the tall bramble surrounding the waste site. We trudged along stealthfully behind her for 20 minutes or so. Finally, she admitted we were lost. The toxic pond, we all believed, lie somewhere, mysteriously, above us.

Diana sat down on a fallen log to rest, tired and short of breath from the fumes of the dump she could feel, but not locate.

Pricked and scratched by brambles, covered with weeds and stickers, we managed to scramble down from the woods—all in about two hours. We wanted to go home.

Driving our rental car out of Yukon, we passed the home of Ben and Bob Ruzina, whom we had met the previous night at a community house meeting. We stopped to say goodbye.

Ben and Bob greeted us at the door.

"789148," said Ben.

"Michigan," said his brother Bob.

They smiled, stepped away from the door jamb, and waved us inside. Ben removed a notepad and pen from the back pocket of his grey work trousers and carefully inscribed the license number of our rented car into his list of outsiders' vehicles.

"If you were one of them," said Bob, "we'd be tracking you."

"We'd keep track," insisted Ben.

In many community organizations, Ben and Bob would qualify as precisely the people to be shunted off during a public moment. Gentle and slow-witted, the two men in their mid-fifties answered most questions with a sort of vague mumble, one brother echoing the other. But CRY made room for everybody who really wanted to work, and Ben and Bob proved tireless campaigners. At the organization's headquarters, the mural of photographs covering the group's "actions" documented the extent of their commitment. Ben and Bob appeared in every march and picket line, standing next to each other, almost touching, arms rigid, gazing straight into the camera with their wan and wary half-smiles. When not attending CRY demonstrations and meetings, or working together as grounds keepers at the nearby golf course, they devoted an intense effort to spotting and logging the license numbers and states of origin for every truck hauling chemical wastes into Yukon.

Ben and Bob Ruzina.

Ben and Bob stared at the small pieces of the forest that still clung to our shirts and pants from our excursion with Diana into the woods earlier that morning. We explained our failed mission. They immediately grabbed us by the arms, and we reluctantly stamped back into the woods behind the Ruzina brothers' house in search, yet again, of the dump.

"I know this place like the back of my hand," assured Ben.

"Back of his hand," agreed Bob.

Within minutes, they penetrated the tangle of brambles and led us to a wind-swept ledge providing a clear, wide view of the dump. We looked at the lake of oily, foul-smelling liquid in a plastic-lined pit about the size of a football field.

"Smells like something dead," said Ben.

"Dead animal," said Bob.

The fumes made our eyes water.

"It's so deep, they could fit three Cincinnati Reds stadiums in there."

"Five stadiums."

"We're not sure," admitted Bob, "but Diana could tell you."

They pushed through the bramble to another spot on the hill and pointed at the dumpsite's main gate.

"That's where the trucks come in."

"Lots of trucks." The brothers recited the names of the states where the trucks originated, and a few of their license numbers.

We knew from our conversations with a number of other Yukon residents that most people viewed the dump as something more than a concrete receptacle of uncertain vast proportions containing hazardous wastes. The dump also stood as a symbolic affront to the community. In towns like Yukon, residents regarded the local leaking toxic waste dump as the sign of a society that no longer dealt with the problems it created. Among the conservative, small-town and rural people who for decades had decried the varied, mounting, and seemingly boundless excesses of American culture—from trillion dollar government debt to the explosive violence of the cities—the uncontainable poisons seemed another object lesson in limits. Beyond the real or imagined threats to public health, the toxic waste dump served as the perfect totem for a society intent on burying its problems or unloading them far from their point of origin for strangers to endure. The leaking ponds were an indictment of technology advanced beyond common sense, a violation of simple decency.

Ben and Bob led us back to the graveyard through a bramble-free shortcut. They galloped down the hill to a tall grey tombstone and assembled respectfully alongside the grave.

"Grandmother," said Ben.

"Our grandmother," echoed Bob.

They moved on to the grave of their brother, who had lived with them until his death from cancer one year before. Bob's chin fell to his chest.

Barely audible, Ben said: "He was too thin."

Bob agreed: "Couldn't gain weight."

As we joined the Ruzina brothers at the family grave, Ben explained that chemicals had caused his brother's cancer. Bob added that the chemicals came from the dump. Yet even Diana Steck had agreed there was no way to prove that the toxic dump had spawned cancers in Yukon. She and the members of CRY knew the town's small population would stymie even the best health studies. When we suggested to Ben and Bob that it would be difficult to prove the source of their brother's illness, they both lowered their heads and mumbled. In truth, our comments were irrelevant.

"He was too thin."

"Couldn't gain weight."

And once again they were off and running, this time to the oldest part of the graveyard.

They halted abruptly beside a row of marble headstones. They faced one grave, their shoulders almost touching, waiting for us. Ben pointed to a small circular tintype photograph embedded in the marble. The ancient photograph portrayed an elderly couple, two of the

Yukon, Pennsylvania.

area's first settlers. The man stared at us through time. But the woman's face was unrecognizable, its border chipped and deeply pitted.

"Kids with guns use the pictures for target practice."

"Only kids."

The Ruzina brothers were close to tears. They stared at the shattered photograph, and then Bob raised his eyes, and Ben did the same. They followed the tree line of the sheltered graveyard up to the hillside that bordered Yukon's toxic dump.

"It's wrong," insisted Ben.

"Just wrong," echoed Bob.

NOTES

Unless otherwise cited, all direct quotes come from personal interviews with the authors. For their time and cooperation, we would like to thank: Terri Swearingen, Connie Rosales, Tina Bravo, Teresa Buentello, Rose Mary Esparza, Borjas Gonzales, Sally Gonzales, Esmeralda Sanchez, Carlos Sanchez, Ed Dunne, Chris Price, Tracey Ramirez, Dolores Huerta, Sister Pat Drydyk, Dr. Lynn Goldman, Rueben Garza, Dr. Beverly Paigen, Martha Salinas, Merry Ellen Alls, Steven Schilling, George Johnson, Dr. Richard Kreutzer, Dr. Marion Moses, Penny Newman, Shirley Ann Neal, Martha Rodriquez, Dr. Raymond Neutra, Dr. Dean Baker, Dr. Elizabeth Whelan, Ramona Franco, Danny Shepherd, Pat Shepherd, Nick Irmiter, Angie Irmiter, Ken McCalip, Patricia Prisbey, Kathy Hoxie, Diana Steck, Dave Rupp, Carol Rupp, Lois Gibbs, Larry Wilson, Sheila Wilson, Gene Hurst, J. D. Boyd, Steve Sterling, Rose Marie Augustine, Linda King, Linda Paxton, Steve Blanchard, Bob Aufiero, Darrell Stinnett, Jim McFarland, Sally Teets, Patty Wallace, Nancy Anderson, Diana Schlaufman, Elaine Drotleff, Lois Kinter, Scott Medwid, Sam Lyle-Medwid, Franklin Rickett, Vern Hurst, Linda Young, Will Collette, Chris Borello, Robert McIntyre, Kari Pratt, Dot Dimitri, Dr. Gerald Ross, Ellen Widess, Mark Mendell, Kaye Kiker, Dr. Ben Chavis, Dr. Robert Bullard, John Zippert, Wendell Paris, Drayton Pruitt, Gordan Kenna, Cleo Askew, Linda Munoz, Charles Munoz, Genaro Zavala, Ben Ruzina, Bob Ruzina.

PROLOGUE

xi "People at the grassroots ...": Albert Gore, "Living Wisely With Creation," *Churchwoman*, 1991, Vol. 6.

xii A local newspaper poll reported ...: *Herald Star* (Steubenville, Ohio), November 18, 1992.

CHAPTER 1. INTO THE HEART OF TOXIC AMERICA

4 The U.S. government estimates ...: The exact amount of toxic materials produced by industry and agriculture, and then released into the environment, remains controversial. For varying estimates see: *Safety On Tap: A Citizen's Drinking Water Handbook*, League of Woman Voters Education Fund, Washington, D.C., 1987, pages 1, 2, 9, 22; "Pesticides in Groundwater: Background Document," Environmental Protection Agency, Washington, D.C., 1986; *Protecting the Nation's Groundwater from Contamination*, Vols. I, II, U.S. Congress, Office of Technology Assessment, Washington, D.C., 1984; Ruth Patrick, Emily Ford, and John Quarles, *Groundwater Contamination in the United States* second edition (Philadelphia: University of Pennsylvania Press, 1987); Samuel S. Epstein, M.D., Lester O. Brown, and Carl Pope, *Hazardous Waste In America* (San Francisco: Sierra Club Books, 1982); John Harte, Cheryl Holdren, Richard Schneider, and Christine Shirley, *Toxics A To Z: A Guide To Everyday Pollution Hazards*, "The Waste Crisis: Sources and Solutions" (Berkeley: University of California Press, 1991). Updated figures can

be obtained from the Toxics Release Inventory User Support Service, Environmental Protection Agency, Washington, D.C.

CHAPTER 2. "THE WHOLE NEIGHBORHOOD WAS STUNNED" —MCFARLAND, FALL 1987

9 "The number of deaths we've had . . . ": Michael McCabe, "Valley Town Feels It's Hexed," *San Francisco Chronicle*, February 18, 1987.

9 "Logic tells us . . . " *Ibid.*

12 On the flip side of her monthly water bill . . . : McFarland water bill, April 1987.

13 "We had nothing but little kids . . . ": Janny Scott, "Child Cancer Cluster Poses Puzzle," *Los Angeles Times*, September 21, 1988.

16 At his funeral . . . : Kathy Freeman, "Parents Say Goodbye to Son, A Victim of Childhood Cancer," *Bakersfield Californian*, July 15, 1987.

18 During the 1960s grape strike . . . : Peter Matthiessen, *Sal Si Puedes: Cesar Chavez and the New American Revolution* (New York: Dell Publishing Company, 1969) page 59. For the early days of the United Farm Workers union, also see: John Gregory Dunne, *Delano* (New York: Farrar, Straus, and Giroux, 1971).

CHAPTER 3. POISONOUS DOUBTS

27 In 1989, following a report . . . : See "How Safe Is Your Food?" *Newsweek*, March 27, 1989; "Is Anything Safe?" *Time*, March 27, 1989; "Dangers in the Vegetable Patch," *Time*, January 30, 1989; Richard L. Berke, "Oratory of Environmentalism Becomes the Sounds of Politics," *New York Times*, April 17, 1990; "Darman Warns Nation of 'Risk-o-Phobia,' " *New York Times*, May 3, 1990.

27 Almost all serious discussion . . . : For the controversy regarding dioxin at Times Beach, see: Eric Felten, "The Times Beach Fiasco," *Insight*, August 12, 1991; "German Study Links Dioxin and Cancer," *New York Times*, October 18, 1991; "The N.Y. Times Detoxifies Dioxin (Again)," *Rachel's Hazardous Waste News*, No. 310, November 4, 1992.

28 In truth, the experts in government . . . : "Researchers Cite Fears Over Toxin Safeguards," *New York Times*, May 17, 1990.

28 A California State law . . . : Elliot Diringer, "Slow Going on Testing of Pesticide Bugs Senator Petris," *San Francisco Chronicle*, March 4, 1987; Sabin Russell, "New State Agency Getting Tough with Pesticide Makers," *San Francisco Chronicle*, August 22, 1991.

28 Today, uncertainty remains the rule . . . : Adam Clymer, "Polls Find Differing Views of Environmental Risks," *New York Times*, May 22, 1989; William K. Stevens, "What Really Threatens the Environment?" *New York Times*, January 29, 1991; William K. Hallman and Abraham H. Wandersman, "Hazardous Waste: Present Risk, Future Risk, Cancer Risk or No Risk?" A paper presented at the American Psychological Association Conference, New Orleans, August 1989; Jane E. Brody, "In Search of Perspective When Fear of Chemicals in Foods Begins to Become a National Problem," *New York Times*, March 23, 1989; Elliot Diringer, "U.S. Awash in Toxics Chemicals—and Fear

of Them," *San Francisco Chronicle*, October 17, 1988; Lydia O. Cunningham, "Living Through a Toxic Accident," *Woman's Day*, November 22, 1988; Stuart Diamond, "Problems at U.S. Plants Raise Fear Over Safety," *New York Times*, November 25, 1985; William Matthews, "U.S. Mood Altered on Toxic Chemicals," *Washington Times*, September 3, 1985.

28 In 1962, Rachel Carson's *Silent Spring* . . . : Rachel Carson, *Silent Spring* (Boston: Houghton Mifflin Company, 1962) page 6.

29 "Exposure to toxic materials not only changes . . . ": For some intriguing analyses of the psychological and neurological effects of contamination, see: Michael R. Edelstein *Contaminated Communities: The Social and Psychological Impacts of Residential Toxic Exposure* (Boulder, Colorado: Westview Press, 1988); Henry M. Vyner, M.D., *Invisible Trauma: The Psychosocial Effects of Invisible Environmental Contamination* (New York: The Free Press, 1988); Robert Jay Lifton, *History and Human Survival* (New York: Random House, 1970); Edward L. Baker, M.D., et al. "Psychiatric Changes in Organophosphate Poisoning: Psychoses and Depression," *South African Medical Journal*, July 2, 1988.

CHAPTER 4. THE MOTHERS' CRUSADE

50 Diana Steck and Lois Gibbs were right to worry . . . : For media coverage regarding this issue, see: April Lynch, "Experts Testify on Toxics' Peril to Children," *San Francisco Chronicle*, September 7, 1990; Philip J. Hilts, "U.S. Opens a Drive on Lead Poisoning in Nation's Young," *New York Times*, December 20, 1990; Natalie Angier, "Study Finds Mysterious Rise in Childhood Cancer Rate," *New York Times*, June 26, 1991; "Heart Study Focuses on Outside Influences: Researchers Say Some Birth Defects Are Caused by Exposure to Toxins Early in Gestation," *San Francisco Chronicle*, April 17, 1992; Mike Magner, "Book Links Pollution, Shifts in Sexuality," *San Francisco Chronicle*, September 20, 1992; Jane E. Brody, "Study Documents Lead-Exposure Damage in Middle-Class Children," *New York Times*, October 29, 1992; Sandra Blakeslee, "Research on Birth Defects Turns to Flaws in Sperm," *New York Times*, January 1, 1991.

51 In Idaho's Silver Valley . . . : Nicholas Freudenberg, *Not In Our Backyards: Community Action for Health and the Environment* (New York: Monthly Review Press, 1984) page 23.

51 In Lowell, Massachusetts . . . : David Ozonoff, "Health Problems Reported by Residents of a Neighborhood Contaminated by a Hazardous Waste Facility," *American Journal of Industrial Medicine*, 1987, 11: 581–597.

51 In Columbia, Mississippi . . . : *Everyone's Backyard*, Citizen's Clearinghouse for Hazardous Wastes, Spring 1988.

51 Beyond the paradoxical attractions of toxic dumps . . . : Lawrie Mott, *Pesticide Alert* (San Francisco: Sierra Club Books, 1987) and personal communication with the authors.

51 Kids' immature digestive tracts . . . : Jane Kay, "Oakland Kids Still Exposed to Lead," *San Francisco Examiner*, December 9, 1990, and personal communication with Lawrie Mott.

52 Since even low levels of lead . . . : *Rachel's Hazardous Waste News* No. 189, July 11, 1990; Beverly Paigen and Lynn Goldman, "Prevalence of Health Problems in Children Living Near Love Canal," *Hazardous Waste and Haz-*

ardous Materials, Vol. 2, No. 1, 1985, pages 23–43; Philip Shenon, "Despite Laws, Water in Schools May Contain Lead, Study Finds," *New York Times,* November 1, 1990.

52 Dr. William Roper . . . : Philip J. Hilts, "U.S. Opens a Drive on Lead Poisoning in Nation's Young—End is Sought in Decade," *New York Times,* December 20, 1990.

52 "We're twenty years behind . . . ": Robert McClure and Linda Kleindienst, "Chemicals Threaten Children," *Florida Sun Sentinel,* December 29, 1991.

52 When the EPA examined . . . : "Human Breast Milk is Contaminated," *Rachel's Hazardous Waste News* No. 193, August 8, 1990.

52 Even prior to breast feeding . . . : Greta G. Fein, et al. "Prenatal Exposure to Polychlorinated Biphenyls: Effects on Birth Size and Gestational Age," *Journal of Pediatrics,* August, 1986, pages 315–319.

53 "You don't have to be Sigmund Freud . . . ": Devra Lee Davis, "Workplace Risks Affect Men, Too," *New York Times,* March 1, 1991.

53 Prior to birth . . . : Sandra Blakeslee, "Research on Birth Defects Turns to Flaws in Sperm," *New York Times,* January 1, 1991; Sandra Blakeslee, "Sometimes It's Dad's Fault: Damaged Sperm Can Cause Congenital Defects in Children," *San Francisco Chronicle,* April 17, 1991.

53 The *Journal of the National Cancer Institute* . . . : Ruth A. Lowengart, et al. "Childhood Leukemia and Parents' Occupational and Home Exposures," *Journal of the National Cancer Institute* Vol. 79, No. 1, July 1987, pages 39–45.

CHAPTER 5. "WHO DO YOU WAVE YOUR SWORD AT?" —MCFARLAND, WINTER 1988

74 To begin, the county team . . . : Leon M. Hebertson, M.D., "McFarland Childhood Cancer Cluster, Phase I," Kern County Department of Health.

75 "We have two choices here . . .": Russell Clemings, "New Fears About Cancer Cluster," *San Francisco Examiner,* September 13, 1987.

75 The video featured . . . : *The Wrath of Grapes,* United Farm Workers Union, 1986.

76 By far the greatest surprise . . . : Russell Clemings, "State Broadens Its Cancer Probe," *Fresno Bee,* September 1, 1987.

77 "We should not let childhood cancer . . .": Elliot Diringer, "State Hearing on Cancer 'Epidemic' in Central Valley," *San Francisco Chronicle,* October 17, 1987.

77 "Some are guilty of the Chicken Little syndrome . . .": Michael Weisskopf, "Pesticides and Death Amid Plenty," *Washington Post,* August 30, 1988.

78 In an editorial . . . : Paul Wahl, "The Politics of Cancer in McFarland," *Record* (Delano), December 3, 1987.

78 As if to confirm . . . : Barry Ginsberg, "Rosales is Harvey's 'Woman of the Year,' " *Record* (Delano), March 1, 1988.

CHAPTER 6. A CHEMISTRY LESSON FOR YELLOW CREEK, KENTUCKY

86 "Carbon is, in fact, a singular element . . . ": Primo Levi, *The Periodic Table* (New York: Schocken Books, 1984).

87 "Organic chemistry to me appears . . . ": Keith Gordon Irwin, *The Romance of Chemistry* (New York: The Viking Press, 1959) page 90.

88 By the late nineteenth century, the benzene ring . . . : For a detailed history of the German chemical industry, see: Peter H. Spitz, *Petrochemicals: The Rise of an Industry* (New York: John Wiley & Sons, Inc., 1988).

88 In fact, the Allied forces . . . : For the critical role of the chemical industry in World War II, see: Lee Niedringhaus Davis, "Stoking the Engines of War" *The Corporate Alchemists: Profit Takers and Problem Makers in the Chemical Industry* (New York: Morrow and Company, 1984).

89 "Not only do chemicals permeate . . . ": *Ibid.* page 278.

89 From 1923 to 1950 . . . : *Ibid.* page 246.

90 Take the case of DDT . . . : John Harte, Cheryl Holdren, Richard Schneider, and Christine Shirley, *Toxics A To Z: A Guide To Everyday Pollution Hazards* (Berkeley: University of California Press, 1991) page 286–288.

91 BCME is a solvent used to manufacture . . . : Robert Gosselin, Roger Smith, and Harold Hodge, *Clinical Toxicology Of Commercial Products,* fifth edition (Baltimore: Williams & Wilkins, 1984) page 183.

91 The most subtle changes in the structure . . . : See Robert Gosselin and Curtis Klassen, Ph.D., et al. *Casarett and Doull's Toxicology* (New York: MacMillan, 1986).

92 The alkali works depended upon manufacture of sulphate . . . : Anthony S. Wohl, *Endangered Lives: Public Health in Victorian Britain* (Cambridge, Massachusetts: Harvard University Press, 1983) page 225. For more information on England's alkali works, see the chapter titled " 'The Black Canopy of Smoke': Atmospheric Pollution."

92 In the words of one historian . . . : *Ibid.* page 226.

93 "One did not have to possess . . . ": *Ibid.* page 264. See the chapter titled " 'The Canker of Industrial Diseases' " for more information about the chemical poisoning of early industrial workers.

94 In Columbia, Mississippi . . . : *Everybody's Backyard,* Citizen's Clearinghouse for Hazardous Wastes, Spring 1988.

94 In Jersey City, New Jersey . . . : Randy Diamond, "New Crisis: Worse Than Love Canal?" *San Francisco Examiner,* January 7, 1990.

95 In Toone, Tennessee . . . : Authors' interviews with residents.

96 In Tucson, Arizona . . . : Authors' interviews with residents. Also see: Jane Kay, "A Deadly Plume Threatens Tucson," *High Country News,* May 25, 1987.

96 In Nitro, West Virginia . . . : The King family's problem was outlined in interviews with the authors. For information on elevated cancer rates along the state's petrochemical corridor, see: Marise S. Gottlieb and Jean K. Carr, "Case-Control Cancer Mortality Study and Chlorination of Drinking Water in Louisiana," *Environmental Health Perspective,* Vol. 46, 1982; Marise S. Gottlieb, "Cancer in Louisiana: An Epidemiological Approach to Exploring Environmental Contributions," *Journal of Environmental Science and Health,* 1983; Marise S. Gottlieb, Jean Carr and Jacquelyn R. Clarkson, "Drinking Water and Cancer in Louisiana," *American Journal of Epidemiology,* Vol. 116, No. 4, 1982; Gottlieb, Carr and Daniel T. Morris, "Cancer and Drinking Water in Louisiana: Colon and Rectum," *International Journal of Epidemiology* Vol. 10, No. 2, 1981.

97 Throughout the country . . . : For an account of several buy-outs, see: Keith Schneider "Safety Fears Prompt Plans to Buy Out Neighbors," *New York Times,* November 28, 1990.

97 In Brookhurst, Wyoming . . . : Authors' interviews with residents.

97 In Ponca City, Oklahoma . . . : Robert Suro, "Refinery's Neighbors Count Sorrows as Well as Riches," *New York Times,* April 4, 1990.

97 In Oxnard, California . . . : Authors' interviews with former residents.

98 Every day in the Mexican town . . . : Ana Arana, "The Wasteland," *Image* in the *San Francisco Examiner,* August 30, 1992.

98 In Rednersville, Ontario . . . : Tom Spears, "Well Contamination: One Community's Legacy of Fear," *Ottawa Citizen,* February 9, 1992; "10 Local Sites That Worry Environmentalists," *Gazette* (Montreal, Canada), October 16, 1991.

99 In Pompano Beach, Florida . . . : Eleanor Smith, "Angry Housewives," *Omni,* December 1986.

99 In Pearland, Texas . . . : *Everybody's Backyard,* Citizen's Clearinghouse for Hazardous Wastes, January 1990, and authors' interview. For more information about SLAPPs (Strategic Lawsuits Against Public Participation), see: "SLAPPing the Opposition" *Newsweek,* March 5, 1990; Jason Zweig, "A SLAPP in the Face," *Forbes,* March 29, 1989; Penelope Canan and George W. Pring, "Strategic Lawsuits Against Public Participation," *Social Problems,* December 1988.

99 In Springfield, Vermont . . . : "After Long Wait, Help for Cleaning a Toxic Dump," *New York Times,* January 2, 1990.

99 "I remember growing up . . . ": In an interview with the authors. For detailed accounts of the Woburn childhood cancer cluster, see: Phil Brown and Edwin J. Mikkelsen, *No Safe Place: Toxic Waste, Leukemia, and Community Action* (Berkeley: University of California Press, 1990) and Paula DiPerna, *Cluster Mystery: Epidemic and the Children of Woburn, Mass.* (St. Louis: The C. V. Mosby Company, 1985). For a comprehensive view of contamination produced by the U.S. Military, see: Seth Shulman, *The Threat At Home: Confronting the Toxic Legacy of the U.S. Military* (Boston: Beacon Press, 1992).

CHAPTER 7. SALLY TEETS AND THE SMALL-TOWN TOXIC REBELS

106 Will Collette was a professional . . . : See: Will Collette, *The Polluters' 'Secret Plan'—And What You Can Do To Mess It Up!* Citizen's Clearinghouse for Hazardous Wastes, Arlington, VA, June 1989.

107 In abundant quantities, she possessed . . . : See: Saul D. Alinsky, *Reveille For Radicals* (New York: Random House, 1969) and Alinsky, *Rules for Radicals* (New York: Random House, 1971). For sympathetic accounts of Alinsky's life and work, see: P. David Finks, *The Radical Vision Of Saul Alinsky* (New York: Paulist Press, 1984) and Sanford D. Horwitt, *Let Them Call Me Rebel: Saul Alinsky, His Life and Legacy* (New York: Alfred A. Knopf, 1989).

108 "If a People's Organization . . . ": Saul D. Alinsky, *Reveille For Radicals,* page 64.

109 To an equal extent . . . : For two excellent accounts, see: Sara M. Evans, *Born For Liberty: A History of Women in America* (New York: The Free Press, 1989) and Anne Firor Scott, *Natural Allies: Women's Associations in American History* (Urbana: University of Illinois Press, 1991).

109 "It also sustained . . . ": Evans, pages 3 and 4.

109 Of all the reform movements driven . . . : See Jack S. Blocker, Jr., *American Temperance Movements: Cycles of Reform* (Boston: Twayne Publishers, 1989) and Ruth Bordin, *Women and Temperance: The Quest for Power and Liberty, 1873–1900* (Philadelphia: Temple University Press, 1981).

110 "In the temperance movement, women could view . . . ": Bordin, page 8.

110 Mary Livermore, a prominent . . . : Quoted in Bordin, page 30.

110 Frances Willard, founder of the Women's Christian . . . : Quoted in Evans, page 128.

110 "Men take one line . . . ": Quoted in Evans, page 129.

110 Willard assured her colleagues . . . : Quoted in Bordin, page 33.

112 When the Sunday *New York Times* . . . : "Editorial Notebook: The Horror Files," *New York Times*, April 12, 1992.

113 But by 1990, people *were* returning . . . : Sam Howe Verhovek, "At Love Canal, Land Rush on a Burial Ground," *New York Times*, July 26, 1990; Michael Winerip, "Home Bargains in Niagara: Just Forget the Toxic Image," *New York Times*, May 29, 1990.

115 "The NIMBY syndrome . . . ": Quoted in Charles Piller, *The Fail-Safe Society: Community Defiance and the End of American Technological Optimism* (New York: Basic Books, 1991) page 4. For industry's view of NIMBYism, also see: "Fighting the NIMBY Syndrome," a special issue of *Waste Age*, March 1988.

115 According to journalist Charles Piller . . . : Piller, page xi.

116 Before leaving California . . . : Fred Setterberg and Lonny Shavelson, "NIMBY Activists Spin Cross-Country Web—Targeting Planet Earth," Pacific News Service, May 8, 1990.

120 "Ecology has become the political substitute . . . ": Quoted in Peter N. Carroll, *It Seemed Like Nothing Happened: The Tragedy and Promise of America in the 1970s* (New York: Holt, Rinehart, Winston, 1982) page 125.

120 The Toxic Substances Control Act . . . : For an insider's account of EPA at the time of this legislation, see Glenn E. Schweitzer, *Borrowed Earth, Borrowed Time: Healing America's Chemical Wounds* (New York: Plenum Press, 1991), particularly pages 10–16.

122 "It is hard to imagine . . . ": Quoted in Jonathan Lash, Katherine Gillman, and David Sheridan, *A Season of Spoils: The Story of the Reagan Administration's Assault on the Environment* (New York: Random House, 1984) page 61. This book is probably the best history of the first Reagan administration's environmental policies.

122 Before heading off . . . : Lash, pages 42–43.

123 Shortly after Love Canal . . . : Schweitzer, page 130.

123 "Abandoned wastes were uncovered . . . ": Schweitzer, page 26.

123 Today Superfund lists . . . : For accounts of Superfund's various woes, see: Environmental Protection Agency, *Environmental Protection Agency's Management of the Superfund Program—An Overview*, December 1988; "Health Risk Assessments at Waste Sites Assailed," *San Francisco Chronicle*, September 4, 1991; Warren E. Leary, "E.P.A. Research Lags, Report Finds," *New York Times*, March 20, 1992; Frank Viviano, "How Superfund Became a Mess," *San Francisco Chronicle*, May 30, 1991; Elliot Diringer, "How Toxic Cleanups Bog Down," *San Francisco Chronicle*, May 31, 1991; Jocelyn White, "Superfund: Pouring Money Down a Hole," *New York Times*, April 17, 1992.

123 According to one of four devastating critiques . . . : *Are We Cleaning Up?: 10 Superfund Case Studies,* Office of Technology Assessment, June 1988.

123 "This is a program that hardly ever . . . ": Frank Viviano, "Superfund Costs May Top S&L Bailout," *San Francisco Chronicle,* May 29, 1991.

123 "Nowhere did I encounter . . . ": Schweitzer, page 134.

124 Some analysts now believe Superfund costs . . . : Schweitzer, Viviano; and Carolyn Lochhead, "Soaring Cost Predicted for Toxic Cleanup," *San Francisco Chronicle,* October 19, 1991.

124 If all this wasn't enough . . . : For the revolving door syndrome, see: Barnaby J. Feder, " 'Mr. Clean' Takes on the Garbage Mess," *New York Times,* March 11, 1990; Jim Sibbison, "Revolving Door at the E.P.A.," *Nation,* November 6, 1989; and Glenn E. Schweitzer, "The Education of the Revolving Door Regulators," pages 18–23.

124 Spurred on by the Reagan administrations' hostility . . . : Peter Borrelli, editor, *Crossroads: Environmental Priorities for the Future* (Washington, D.C.: Island Press, 1988) page 10.

125 By 1992, the president of the National Wildlife . . . : Keith Schneider, *New York Times,* March 29, 1992.

125 "The Sierra Club is becoming . . . ": Quoted in Christopher Manes, *Green Rage: Radical Environmentalism and the Unmaking of Civilization* (New York: Little, Brown, and Company, 1990) page 45.

126 "Pollution, defilement, equalor . . . ": Quoted in Borrelli, page 73.

127 Not only had the liberal . . . : Christopher Lasch, *The True and Only Heaven: Progress and Its Critics* (New York: W. W. Norton and Company, 1991) page 17.

128 According to Cerrell . . . : "Political Difficulties Facing Waste-To-Energy Conversion Plant Siting," Cerrell Associates, Los Angeles, CA, 1984.

CHAPTER 8. "IT'S LIKE A MURDER MYSTERY: WE'VE ELIMINATED THE BUTLER—BUT WHO DID IT?" —MCFARLAND, SPRING 1988

131 For a summary and analyses of the study, see: California Department of Health Services, Environmental Epidemiology and Toxicology Program, "Summary of Environmental Data: McFarland Childhood Cancer Cluster Investigation," Phase III Report, October 1991; California Department of Health Services, Environmental Epidemiology and Toxicology Program, *McFarland Community Health Newsletter,* Vol. 1, No. 2, October 1, 1988; California Department of Health Services, *History of the McFarland Health Studies,* October 1991; California Department of Health Services, "Questions and Answers about the McFarland Cancer Cluster," October 1990.

133 Researchers collected soil samples . . . : Lockheed Engineering and Sciences Company, "X-Ray Fluorescence Site Screening Survey, Soil Sampling, and Chemical Analyses, McFarland, California," Las Vegas, Nevada, Office of Research and Development, U.S. Environmental Protection Agency, December 1990.

133 The level of doubt . . . : Louis Slesin, "Making Waves in McFarland," *Bakersfield Californian,* March 17, 1988; Slesin, "California Cancer Cluster: Is RF Radiation Involved?" *Microwave News,* January/February 1988.

134 The Department of Health Services research staff . . . : California Department of Health Services, Environmental Epidemiology and Toxicology Program, "Summary of Environmental Data: McFarland Childhood Cancer Cluster Investigation" Phase III Report, October 1991, page 73.

135 But these reports had one major flaw . . . : California Department of Health Services, Environmental Epidemiology and Toxicology Program, "Summary of Environmental Data: McFarland Childhood Cancer Cluster Investigation" Phase III Report, October 1991, page 85. For more on pesticides, see: Bob Secter, "Harvesting Limits on Chemicals," *Los Angeles Times,* May 23, 1990; Jack Anderson, "EPA's Pesticide Testing Lost in a Bureaucratic Thicket," *San Francisco Chronicle,* March 19, 1990; Marion Moses, M.D., "Pesticides Plague Farmworkers, Consumers," *National Catholic Rural Life,* February 1988; John T. O'Connor and Sanford Lewis, *Shadow on the Land,* National Toxics Campaign, Boston, 1988; Peter Weber, "A Place for Pesticides?" *World Watch,* May/June 1992.

135 Not long after the state released . . . : Ray Sotero, "Cancer-Causing Pesticide Spread By Fog, Study Says," *San Francisco Examiner,* December 17, 1989.

136 Esparza's comments pointed directly . . . : For coverage of environmental health studies, see: Elliot Diringer, "Experts Struggle with Toxics' Role in Cancer," *San Francisco Chronicle,* October 19, 1988; Penny Newman, "Cancer Clusters Among Children: The Implications of McFarland," *Journal of Pesticide Reform,* Vol. 9, No. 3, Fall 1989; "Fresno Study Finds No Link in Child Leukemia Victims," *San Francisco Chronicle,* May 20, 1987; Joyce Price, "Study Links Childhood Leukemia to Use to of Pesticides by Parents," *Washington Times,* July 13, 1987.

137 "My interest, number one, is for the families . . . ": Kathy Freeman, "Growers React to State Report with Uncertainty," *Bakersfield Californian,* January 30, 1988. This article also contains the quotes from Don Riley, Jack Pandol, Sr., and Dolores Huerta.

138 "You hope that you find a smoking gun . . . ": Kathy Freeman, "State: Still No 'Smoking Gun' in McFarland Cancer Mystery," *Bakersfield Californian,* July 23, 1987.

139 "I saw a lady with a very small . . . ": Sally Connell, "Potent Adult Cancer Linked to McFarland," *Bakersfield Californian,* May 19, 1988.

140 Surprisingly, they found the highest incidence . . . : Nancy Weaver, "Town's Kids Suffer High Cancer Rate," *San Francisco Chronicle,* April 5, 1987; Louis Sahagun, "Rosamond: Malignant Mystery," *Los Angeles Times,* September 30, 1988; Sahagun, "No Cancer Link Found in Tests of 22 Sites at Rosamond," *Los Angeles Times,* October 25, 1988.

140 "I have never seen this much waste . . . ": Sahagun, *Los Angeles Times,* September 30, 1988.

140 "The way science works . . . ": Janny Scott, "Child Cancer Cluster Poses Puzzle," *Los Angeles Times,* September 21, 1988.

CHAPTER 9. THE STRUGGLE FOR CERTAINTY: SCIENCE RUNS UP AGAINST ITS LIMITS IN TOXIC TOWNS

143 The Stringfellow Acid Pits contained . . . : Michael Brown, *Laying Waste* (New York: Pocket Books, 1981) page 201; Dean Baker, M.D., M.P.H.,

Sander Greenland, M.S., Dr. P.H., James Mendlein, Ph.D., Patricia Harmon, M.P.H., "A Health Study of Two Communities Near the Stringfellow Waste Disposal Site," *Archives of Environmental Health,* September/October 1988, Vol. 34 No. 5; Michael Weisskopf, "California's Subterranean Toxic Blob: The Stringfellow Acid Pits Legacy Persists," *San Francisco Sunday Chronicle and Examiner, This World,* March 15, 1987.

143 When heavy rains hit . . . : Dean Baker, M.D., M.P.H., and Sander Greenland, M.S., Dr.P.H., "Stringfellow Health Effects Study: An Epidemiologic Health Survey of Residents of Glen Avon and Rubidoux, California," 1986, School of Public Health, University of California, Los Angeles, pages 1.4 and 1.5.

144 Epidemiology's first hero . . . : Elizabeth Whelan, *Toxic Terror* (Ottawa, Illinois: Jameson Books, 1985) page 218.

144 Glen Avon's saga actually begins . . . : Baker et al. "An Epidemiological Health Survey of Residents of Glen Avon and Rubidoux, California," pages 1.4 and 1.5; Klein, Wegis, and Duggan, Attorneys at Law, "Fourth Consolidated Amended Complaints for Money Damages," September 24, 1987, page 29.

145 In 1969, torrential downpours . . . : Baker et al. "An Epidemiological Health Survey of Residents of Glen Avon and Rubidoux, California," page 1.4.

145 Three years later . . . : Baker et al. page 1.6; Klein, Wegis, and Duggan, page 39.

145 The more adventurous kids . . . : Klein, Wegis, and Duggan, page 35.

145 After investigators located the contaminated well . . . : Baker et al. "An Epidemiological Health Survey of Residents of Glen Avon and Rubidoux, California," page 1.4.

145 Heavy rains throughout 1978 . . . : Weisskopf.

146 But shortly after this time, some parents noticed . . . : Klein, Wegis, and Duggan, page 45.

146 The next year, the California State Department of Health . . . : *Ibid.* page 48.

148 Its members included both Penny Newman and Dr. Beverly Paigen . . . : Beverly Paigen and Lynn Goldman, "Prevalence of Health Problems in Children Living Near Love Canal," *Hazardous Waste and Hazardous Materials,* Vol. 2 No. 1, 1985, pages 23–43.

148 In 1983, twelve trained interviewers . . . : Baker et al. "A Health Study of Two Communities Near the Stringfellow Waste Disposal Site," page 325.

149 The state health officials reported . . . : *Ibid.* pages 325–334.

153 According to Dr. Dean Baker and Dr. Sander Greenland . . . : *Ibid.* page 331.

153 "Every time we use an imprecise measure . . . ": Dr. Lynn Goldman, an address to a conference on "Communities and the Environment," Oakland, California, March 22, 1991.

156 The Glen Avon study did come up with . . . : Baker et al. "A Health Study of Two Communities Near the Stringfellow Waste Disposal Site," page 325.

157 The Glen Avon researchers had drawn upon the example . . . : David Ozonoff, et al. "Health Problems Reported by Residents of a Neighbor-

hood Contaminated by a Hazardous Waste Facility," *American Industrial Medicine*, 11:581–97; David Ozonoff et al. "Silresim Area Health Study, Report of Findings," Boston School of Public Health and Center for Survey Research, University of Massachusetts, 1983; Ozonoff and Leslie I. Boden, "Truth and Consequences: Health Agency Responses to Environmental Health Problems," *Science, Technology, & Human Values*, Summer/Fall 1987.

157 As a result, Baker and Greenland modified Ozonoff's approach . . . : Baker et al. "A Health Study of Two Communities Near the Stringfellow Waste Disposal Site," page 329.

158 "For example," their study stated, "it is reasonable . . . : *Ibid.* page 332.

159 Industrial workers exposed to asbestos . . . : Paul Brodeur, *Expendable Americans* (New York: Viking Press, 1974) page 53.

159 The same held true for a rare liver tumor . . . : *Ibid.* pages 251–274.

164 All these problems made it even more surprising . . . : National Research Council, Committee on Environmental Epidemiology, *Environmental Epidemiology: Public Health and Hazardous Wastes* (Washington, D.C.: National Academy Press, 1991).

165 "All this data suggests . . . ": Cynthia Hubert, "Heart Defect Study Cites TCE," *Arizona Daily Star*, September 6, 1987; Stanley Goldberg, "An Association of Human Congenital Cardiac Malfunctions and Drinking Water Contaminants," *Journal of the American College of Cardiology* Vol. 16 No. 1, July 1990, pages 155–164.

166 In San Jose, California . . . : National Research Council, page 133; Epidemiological Studies Section, California Department of Health Services, "Pregnancy Outcomes in Santa Clara County 1980–1982: Two Epidemiological Studies," Berkeley, California, 1985; Arthur C. Upton et al. "Public Health Aspects of Toxic Chemical Disposal Sites," *Annual Review of Public Health* Vol. 10, 1989, page 19.

166 In 1990, Dr. B. V. Dawson . . . : National Research Council, page 140; Brenda V. Dawson et al. "Cardiac Teratogenesis of Trichloroethylene and Dichloroethylene in a Mammalian Model," *Journal of the American College of Cardiology* Vol. 16 No. 5, November 1, 1990, pages 1,304–9.

167 "If we waited for 100 percent . . . ": Elliot Diringer, "Scientist Questions Way Cancer Risk is Assessed," *San Francisco Chronicle*, August 31, 1990.

167 "There are forty million people out there . . . ": Keith Schneider, "U.S. is Said to Lack Data on Toxic Threat," *New York Times*, October 22, 1991.

169 According to a General Accounting Office survey . . . : "Health Risk Assessments at Waste Sites Assailed," *San Francisco Chronicle*, September 4, 1991.

CHAPTER 10. LOVE'S UNCERTAIN CHEMISTRY: THE TRAGEDY AND ROMANCE OF ENVIRONMENTAL ILLNESS IN THE WEST TEXAS DESERT

177 In 1989, the National Foundation for the Chemically . . . : *Cheers*, National Foundation for the Chemically Hypersensitive, Wrightsville Beach, North Carolina, 1989.

177 The National Academy of Sciences approximates . . . : Peter Montague, *Rachel's Hazardous Waste News* No. 220; Environmental Research Foundation, Washington, D.C., February 13, 1991; "Workshop on Health Risks from Exposure to Common Indoor Household Products in Allergic or Chemically Diseased Persons," Board of Environmental Studies and Toxicology, National Research Council, July 1, 1987.

178 "The TB sufferer was a dropout . . . ": Susan Sontag, *Illness As Metaphor* (New York: Vintage Books) page 32.

178 A psychiatrist of today . . . : Eileen Stewart and Joel Raskin, "Psychiatric Assessment of Patients with Twentieth Century Disease (Total Allergy Syndrome)," *Canadian Medical Association Journal* Vol. 133, November 15, 1985, page 1001.

179 Doctors in Toronto, Canada . . . : *Ibid.*, page 1002.

179 Dr. Carroll Brodsky . . . : Carroll Brodsky, M.D., Ph.D., " 'Allergic to Everything': A Medical Subculture," *Psychosomatics* Vol. 24 No. 8, August 1983, page 735.

179 Writing in the *Journal of the American Medical Association* . . . : Study cited in Natalie Angier, "Environmental Illness May Be Mental," *New York Times*, December 26, 1990, and subsequent interview with authors.

180 "I should like to die of consumption . . . ": Quoted in Sontag, page 30.

180 Even if environmental illness does not exist . . . : Sontag, page 33.

181 About this time, Pratt's mother . . . : Elliot Diringer, "U.S. Awash in Toxic Chemicals—and Fear of Them," *San Francisco Chronicle*, October 17, 1990.

181 When Dr. Theron Randolph proposed . . . : For Randolph's theories, see: Theron G. Randolph, M.D., *Human Ecology and Susceptibility To The Chemical Environment* (Springfield, Illinois: Charles C. Thomas, Publishers, 1962).

182 In modern medicine, an allergic reaction . . . : Nicholas A. Ashford and Claudia S. Miller, *Chemical Exposures: Low Levels and High Stakes* (New York: Van Nostrand Reinhold, 1991) pages 20 and 32.

182 Randolph claimed that allergies . . . : Theron G. Randolph, M.D., and Ralph W. Moss, Ph.D., *An Alternative Approach To Allergies* (New York: Lippincott and Crowell, 1980). Other interpretations of environmental illness strongly influenced by Randolph: Richard Mackarness, *Living Safely In A Polluted World* (New York: Stein and Day, 1981) and Iris R. Bell, M.D., Ph.D., *Clinical Ecology: A New Medical Approach to Environmental Illness* (Bolinas, California: Common Knowledge Press, 1982).

183 In the words of one of Randolph's disciples . . . : Mackarness, page 123.

183 "The theory of clinical ecology . . . ": Mackarness, page 172.

184 In their 1991 survey . . . : Ashford and Miller, pages 21, 32, 22, and 60.

184 "Allergists continue to point . . . ": *Ibid.* page 21.

188 Dr. Ephraim Kahn, a toxicologist . . . : Ephraim Kahn, "Chemical Sensitivity: Fact or Fad," *HES Tox-Epi Review*, State of California Department of Health Services, Berkeley, May 1987.

191 Although they didn't know it . . . : Roy Bragg, "Nowhere Left to Go: Chemically Sensitive People Make a Stand," *Texas: Houston Chronicle Magazine*, November 19, 1989; Suzanne Gamboa, "Vineyard Bears Grapes of

Wrath," *Odessa American,* July 1, 1990; Joe Nick Patoski, "Nowhere to Run," *Texas Monthly,* February 1991.

193 Soon after Kari sent her letter . . . : "Davis Mountain Residents Fight for their Lives," *Jefferson Davis County Mountain Dispatch,* January 31, 1991.

194 In 1990, the National Research Council . . . : Mark Mendell, M.P.H., presentation to the California Department of Health Services, July 3, 1991, and "All Things Considered," National Public Radio, October 1, 1991.

CHAPTER 11. "AROUND HERE, YOU'RE EITHER ON ONE SIDE OR THE OTHER!"—MCFARLAND, SUMMER 1988

202 "They really don't care about the children . . . ": Richard Steven Street, "Care Packaging," *California Farmer,* July 16, 1988, page 16.

202 The 1980s had worn hard . . . : For various accounts of the United Farm Workers union's problems, see: Harry Bernstein, "Farm Union: Why Didn't It Burgeon?" *Los Angeles Times,* October 25, 1981; Paul Shinoff, "Wither the Eagle?" *San Francisco Sunday Examiner & Chronicle,* April 8, 1984; Jeff Coplon, "Cesar Chavez's Fall from Grace, Part I & II," *Village Voice,* August 14 & August 21, 1984; Tim Redmond, "The Bay Guardian Interview: Cesar Cheavez" *San Francisco Bay Guardian,* June 26, 1985; Jerome Cohen, "UFW Must Get Back to Organizing," *Los Angeles Times,* January 15, 1986; Harry Bernstein, "Ruling May Devastate Chavez's Union," *Los Angeles Times,* February 25, 1987; Anthony Marquez, "After 25 Years, Chavez Fights Predictions of Union's Demise," *San Jose Mercury News,* July 28, 1987; Anthony Marquez, "Farm Workers Face Dilemma," *San Francisco Examiner,* August 2, 1987; T. T. Nhu, "Chavez Fights On, But Crusade Has Lost Its Force," *San Jose Mercury News,* August 19, 1987.

203 According to the *Wall Street Journal* . . . : Constanza Montana, and John Emshwiller, "Its Ranks Eroding, Farm Workers' Union Struggles to Survive," *Wall Street Journal,* September 9, 1986; also see Delia M. Rios and Dale Rodebaugh, "Chavez's UFW is Fighting for Its Life," *San Jose Mercury News,* May 21, 1987.

203 While the *New Republic* was still characterizing Chavez. . . . : Evan T. Barr, "Sour Grapes," *New Republic,* November 25, 1985.

204 But the second time around . . . : Julia Benitez. "UFW Renews Grape Boycott," *Record* (Delano), July 12, 1984; Mervin Field, "Chavez More Popular Than His Boycott," *San Francisco Chronicle,* September 6, 1985; N. Craig Smith, "California Grapes: A Vintage Boycott," *Business and Society Review,* Summer 1991.

204 "What do you and a farm worker . . . ": Undated direct mail donor solicitation on UFW letterhead.

204 In the spirit of the self-absorbed 1980s . . . : Steve Wiegand, "Cesar Chavez Goes High-Tech," *California Journal,* April 1985.

204 In some fund-raising letters . . . : Undated UFW direct mail donor letter on letterhead of "Bernice Bonillas, Bakersfield, California."

204 In June 1989, a California judge . . . : Mark Hagar, "Legal Lynching: Farm Workers and the Laws," *Z Magazine,* September 1991.

205 "It is much too toxic ..." : "Parents Upset by Ruling," *Record,* September 18, 1986.

206 Yet some inconstant efforts ... : "McFarland Group Airs Concerns," *Record* (Delano), November 11, 1988. Also see undated McFarland United Community flyer outlining the meeting's agenda.

206 "It was so exciting ... ": Sally Connell, "Jackson Lays Blame on Federal Officials," *Bakersfield Californian,* May 25, 1988.

207 "The idea that there is contamination ... ": *Ibid.*

207 Alongside Jackson stood ... : For more about the first Jackson visit to McFarland, see: Sally Connell, "Jackson to Visit McFarland," *Bakersfield Californian,* May 23, 1988; Sally Connell, "Candidate Calls on Couple Whose Son Died of Cancer," *Bakersfield Californian,* May 24, 1988; Scott Forter and Sally Connell, "Presidential Hopeful Rallies Residents, Questions EPA," *Bakersfield Californian,* May 24, 1988; Sally Connell, "Jackson Seeks Cancer Probe," *Bakersfield Californian,* May 25, 1988; Ron Harris, "Jackson to Put Campaign Focus on Cancer Cluster Town," *Los Angeles Times,* May 28, 1988.

208 "Why are we marching ... ": Sally Connell and Scott Forter, "Jackson Rallies Town Hit By Cancer Cluster," *Bakersfield Californian,* June 5, 1988.

209 Nevertheless, he waved his hands ... : Martha Ginsberg, "Jackson Makes Big Pitch Against Killer Pesticides," *San Francisco Examiner,* June 5, 1988.

210 "We do extensive testing ... ": "Grapes are the Safest Thing You Could Eat," *Record* (Delano), October 18, 1988.

210 Throughout this busy political season ... : For information on Cesar Chavez's "Fast for Life," see: Pat Hoffman, "Cesar Chavez's 'Fast for Life,' " *Christian Century,* October 12, 1986; Robert B. Gunninson, "UFW Unrelenting On Table Grapes," *San Francisco Chronicle,* August 8, 1988; Robert B. Gunnison, "Actors Rally Around Chavez," *San Francisco Chronicle,* August 9, 1988; L. A. Chung, "Chavez Ends Fast Amid Celebrities," *San Francisco Chronicle,* August 22, 1988; Richard Rodriquez, "Chavez: The Farmworkers' Prophet," *Contra Costa Times,* August 28, 1988.

210 "The day will come ... ": Louis Sahagun, "3 of Robert Kennedy's Children Visit Chavez to Support Protest," *Los Angeles Times,* August 5, 1988.

210 Chavez referred to his fast ... : "Cesar Chavez Fasts For Life," *National Farm Worker Ministry Newsletter,* Summer 1988.

211 "Our Creator creates scientists ... ": Janny Scott, "Child Cancer Cluster Poses Puzzle," *Los Angeles Times,* September 21, 1988.

211 In December 1988, the McFarland Community Health Center ... : For information on the clinic, see: "Governor Vetoes Bill for McFarland Clinic," *Record* (Delano), October 4, 1988; Sally Connell, "Clinic, built 'the American Way,' Gives McFarland Hope," *Bakersfield Californian;* "Clinic Gets Pandol's $5,000," *Record* (Delano), December 13, 1988.

212 On January 11, 1989, Dr. Ken Kizer ... : Letter from Department of Health Services, dated January 11, 1989.

212 The entire affair harkened back to ... : "Cancer Screening Talk," *Record* (Delano), November 18, 1988; "Health Screening Set in McFarland," *Record* (Delano), January 17, 1989; "Editorial: State's Method of Informing is Lousy," *Record* (Delano), January 18, 1989.

213 The *Record* joined in the confusion ... : *Record* (Delano), January 18, 1989.

215 Fifteen miles north of McFarland . . . : "Chavez's New Plea to Ban Pesticides," *San Francisco Chronicle,* September 15, 1989; Elliot Diringer, "New Cancer Cluster in Farm Town," *San Francisco Chronicle,* September 14, 1988; United Farm Workers direct mail solicitation, dated May 1992; Jane Kay, "Farm Workers Call It 'Toxic Racism,' " *San Francisco Chronicle;* Sally Ann Connell, "And the Children Keep On Dying," *San Francisco Chronicle & Examiner, Sunday Punch,* June 10, 1990; Marion Moses, M.D., "Childhood Cancer in Earlimart," September 14, 1989.

CHAPTER 12. WHO DO YOU TRUST? THE RIDDLES OF ENVIRONMENTAL RACISM IN EMELLE, ALABAMA

218 In 1982, North Carolina governor . . . : Karl Grossman, "From Toxic Racism to Environmental Justice," *E: The Environmental Magazine,* May/June 1992, and Dick Russell, "Environmental Racism," *Amicus Journal,* Spring 1989.

220 In a national survey . . . : "Toxic Wastes and Race in the United States: A National Report on the Racial and Socio-Economic Characteristics of Communities with Hazardous Waste Sites," Commission for Racial Justice, United Church of Christ, Cleveland, Ohio, 1987, page 9.

221 Armed with this incendiary analysis . . . : For further discussion of the environmental movement in minority communities, see: Dana Alston, *We Speak For Ourselves: Social Justice, Race and Environment,* The Panos Institute, December 1990; *Toxic Wastes and Race in the United States: A National Report on the Racial and Socio-Economic Characteristics of Communities with Hazardous Waste Sites,* Committee for Racial Justice, United Church of Christ, Cleveland, Ohio, 1987; *Toxics and Minority Communities,* The Alternative Policy Institute of the Center for Third World Organizing, July 1986; Donald Elliot, Jr., "Environmental Equity in the 1990s: Pollution, Poverty and Political Empowerment," *Kansas Journal of Law and Public Policy* Vol. 1 No. 1 1991; Jane Kay, "Fighting Toxic Racism," *San Francisco Examiner,* April 7, 1991.

221 Bullard termed Emelle's politics . . . : Robert D. Bullard, *Dumping in Dixie: Race, Class and Environmental Quality,* (Boulder, Colorado: Westview Press, 1990) page 73.

224 In the *International Journal of Epidemiology* . . . : Elisabeth, Rosenthal, "Health Problems of Inner City Poor Reach Crisis Point," *New York Times,* December 24, 1990.

224 "We looked at conditions. . . . ": *Ibid.*

224 According to the *New England Journal of Medicine* . . . : *Ibid.*

234 According to a survey. . . . : "Press Reports About Violations of Law by Waste Management, Inc.," Environmental Research Foundation, Washington, D.C., July 1988.

234 In fact, over the years. . . . : "Waste Management Plan to Settle Suit Gets Preliminary Approval," *Wall Street Journal,* May 29, 1985.

234 In 1987, a lobbyist . . . : "U.S. Indicts Seven in Chicago Probe of City Corruption," *Wall Street Journal,* November 24, 1986; Susan Kuczka, United Press International, untitled wire release.

234 In Niagra Falls . . . : Robert McClure and Fred Schulte, "Laws Fail to Prevent Pollution at Dump Sites," *News & Sun Sentinel* (Ft. Lauderdale), December 6, 1987.

243 "Environmental justice demands . . . ": *Urban Ecologist,* Berkeley, Spring 1992.

CHAPTER 13. "TONIGHT FEELS SOMETHING LIKE THE FALL OF SAIGON" —MCFARLAND TODAY

260 The next day . . . : Tom Maurer, "Cancer Clues Elusive: State Abandons Study of McFarland Cluster," *Bakersfield Californian,* October 25, 1991.

261 Dona Amparo, the activist . . . : "Notes from the Playwright," *Heroes and Saints,* Cherrie Moraga.

CHAPTER 14. THE VIRTUES OF MESSY DEMOCRACY AND UNCERTAIN SCIENCE

265 "No one can contemplate . . . ": William Greider, *Who Will Tell The People: The Betrayal of American Democracy* (New York: Simon and Schuster, 1992) pages 406 and 163.

265 But rehabilitating American democracy . . . : *Ibid.* page 162.

266 Still, as we traveled throughout . . . : Keith Schneider, "Pollution From Toxic Chemicals Shows Decline in U.S. Since 1987," *New York Times,* May 17, 1991.

NATIONAL ORGANIZATIONS ADDRESSING HAZARDOUS WASTE ISSUES

Citizen's Clearinghouse for Hazardous
  Wastes
119 Rowell Court
Falls Church, VA 22046

Citizens for a Better Environment
501 Second Street
Suite 305
San Francisco, CA 94107

Clean Sites, Inc.
1199 North Fairfax Street
Suite 400
Alexandria, VA 22314

Earth Island Institute
300 Broadway
San Francisco, CA 94133

Environmental Health Network
P.O. Box 1628
Harvey, LA 70058

Environmental Research Foundation
P.O. Box 73700
Washington, D.C. 20056

Greenpeace
1436 U Street NW
Washington, D.C. 20009

Highlander Center
Route 3, Box 370
New Market, TN 37820

National Coalition Against the Misuse
  of Pesticides
701 E Street, Suite 200
Washington, D.C. 20003

National Toxics Campaign
1168 Commonwealth Avenue
Boston, MA 02134

U.S. Public Interest Research Group
215 Pennsylvania Avenue SE
Washington, D.C. 20003

SELECTED BIBLIOGRAPHY

BOOKS

Alinsky, Saul D. *Reveille for Radicals*. New York: Random House, 1969.
———. *Rules for Radicals*. New York: Random House, 1971.
Alston, Dana. *We Speak for Ourselves: Social Justice, Race and Environment*. The Panos Institute, December 1990.
Andelman, Julian B., and Dwight W. Underhill, eds. *Health Effects from Hazardous Waste Sites*. Chelsea, Michigan: Lewis Publishers, 1987.
Ashford, Nicholas, and Claudia Miller. *Chemical Exposures: Low Levels and High Stakes*. New York: Van Nostrand Reinhold, 1991.
Bell, Iris. R., M.D., Ph.D. *Clinical Ecology: A New Medical Approach to Environmental Illness*. Bolinas, California: Common Knowledge Press, 1982.
Blocker, Jack S., Jr. *American Temperance Movements: Cycles of Reform*. Boston: Twayne Publishers, 1989.
Bordin, Ruth. *Women and Temperance: The Quest for Power and Liberty, 1873–1900*. Philadelphia: Temple University Press, 1981.
Borrelli, Peter. *Crossroads: Environmental Priorities for the Future*. Washington, D.C.: Island Press, 1988.
Brodeur, Paul. *Outrageous Misconduct: The Asbestos Industry on Trial*. New York: Pantheon Books, 1985.
Brown, Michael. *Laying Waste: The Poisoning of America by Toxic Chemicals*. New York: Pocket Books, 1981.
Brown, Phil, and Edwin J. Mikkelsen. *No Safe Place: Toxic Waste, Leukemia, and Community Action*. Berkeley: University of California Press, 1990.
Bullard, Robert D. *Dumping in Dixie: Race, Class, and Environmental Quality*. Boulder, Colorado: Westview Press, 1990.
Carroll, Peter N. *It Seemed Like Nothing Happened: The Tragedy and Promise of America in the 1970s*. New York: Holt, Rinehart and Winston, 1982.
Carson, Rachel. *Silent Spring*. Boston: Houghton Mifflin Company, 1962.
Chavkin, Wendy, M.D., ed. *Double Exposure: Women's Health Hazards on the Job and at Home*. New York: Monthly Review Press, 1984.
Cohen, Jerry, and John O'Connor. *Fighting Toxics: A Manual for Protecting Your Family, Community, and Workplace*. Washington, D.C.: Island Press, 1990.
Commoner, Barry. *Making Peace with the Planet*. New York: Pantheon Books, 1990.
Davis, Lee Niedringhaus. *The Corporate Alchemists: Profit Takers and Problem Makers in the Chemical Industry*. New York: William Morrow and Company, 1984.
De Lillo, Don. *White Noise*. New York: Penguin Books, 1986.
DiPerna, Paula. *Cluster Mystery: Epidemic and the Children of Woburn, Mass*. St. Louis: The C.V. Mosby Company, 1985.
Dunne, John Gregory. *Delano*. New York: Farrar, Straus, and Giroux, 1971.
Edelstein, Michael R. *Contaminated Communities: The Social and Psychological Impacts of Residential Toxic Exposure*. Boulder, Colorado: Westview Press, 1988.

Epstein, Samuel S., M.D., Lester O. Brown, and Carl Pope. *Hazardous Waste in America*. San Francisco: Sierra Club Books, 1982.

Evans, Sara M. *Born for Liberty: A History of Women in America*. New York: The Free Press, 1989.

Feshbach, Murray, and Alfred Friendly, Jr. *Ecocide in the USSR: Health and Nature Under Siege*. New York: Basic Books, 1992.

Finks, P. David. *The Radical Vision of Saul Alinsky*. New York: Paulist Press, 1984.

Freudenberg, Nicholas. *Not in Our Backyards: Community Action for Health and the Environment*. New York: Monthly Review Press, 1984.

Gibbs, Lois. *Love Canal: My Story*. Albany: State University of New York Press, 1982.

Glendinning, Chellis. *When Technology Wounds: The Human Consequences of Progress*. New York: William Morrow and Company, 1990.

Gosselin, Robert, Roger Smith, and Harold Hodge. *Clinical Toxicology of Commercial Products,* 5th ed. Baltimore: Williams & Wilkins, 1984.

Gottlieb, Robert. *A Life of Its Own: The Politics and Power of Water*. San Diego: Harcourt Brace Jovanovich, Publishers, 1988.

Greider, William. *Who Will Tell the People: The Betrayal of American Democracy*. New York: Simon and Schuster, 1992.

Hall, Bob. *Environmental Politics: Lessons from the Grassroots*. Durham, North Carolina: Institute for Southern Studies, 1988.

Harte, John, Cheryl Holdren, Richard Schneider, and Christine Shirley. *Toxics A to Z: A Guide to Everyday Pollution Hazards*. Berkeley: University of California Press, 1991.

Highland, Joseph H. *Malignant Neglect*. New York: Alfred A. Knopf, 1979.

Horwitt, Sanford D. *Let Them Call Me Rebel: Saul Alinsky, His Life and Legacy*. New York: Alfred A. Knopf, 1989.

Irwin, Keith Gordon. *The Romance of Chemistry*. New York: The Viking Press, 1959.

Johnson, Haynes. *Sleepwalking Through History: America in the Reagan Years*. New York: Doubleday, 1991.

Klassen, Curtis, Ph.D., et al. *Casarett and Doull's Toxicology*. New York: MacMillan, 1986.

Krimsky, Sheldon, and Alonzo Plough. *Environmental Exposures: Communicating Risk as a Social Process*. New York: Auburn House, 1988.

Lasch, Christopher. *The True and Only Heaven: Progress and Its Critics*. New York: W.W. Norton and Company, 1991.

Lash, Jonathan, Katherine Gillman, and David Sheridan. *A Season of Spoils: The Story of the Reagan Administration's Assault on the Environment*. New York: Random House, 1984.

Legator, Marvin S., Barbara L. Harper, and Michael J. Scott. *The Health Detective's Handbook: A Guide to the Investigation of Environmental Health Hazards*. Baltimore: The Johns Hopkins University Press, 1985.

Lappé, Marc. *Chemical Deception: Exposing Ten Myths That Endanger Us All*. San Francisco: Sierra Club Books, 1991.

Lester, James P., and Ann O'M Bowman, editors. *The Politics of Hazardous Waste Management*. Durham, North Carolina: Duke University Press, 1983.

Lewis, H.W. *Technological Risk*. New York: W.W. Norton and Company, 1990.

Mackarness, Richard. *Living Safely in a Polluted World.* New York: Stein and Day, 1981.

Manes, Christopher. *Green Rage: Radical Environmentalism and the Unmaking of Civilization.* New York: Little, Brown, and Company, 1990.

Mann, Thomas. *The Magic Mountain.* New York: Alfred A. Knopf, 1934.

Matthiessen, Peter. *Sal Si Puedes: Cesar Chavez and the New American Revolution.* New York: Dell Publishing Company, 1969.

Mokhiber, Russell. *Corporate Crime and Violence: Big Business, Power and the Abuse of the Public Trust.* San Francisco: Sierra Club Books, 1988.

Mott, Lawrie. *Pesticide Alert.* San Francisco: Sierra Club Books, 1987.

National Research Council, Committee on Chemical Toxicity and Aging. *Aging in Today's Environment.* Washington, D.C.: National Academy Press, 1987.

National Research Council, Committee on Environmental Epidemiology. *Environmental Epidemiology: Public Health and Hazardous Wastes.* Washington, D.C.: National Academy Press, 1991.

Patrick, Ruth, Emily Ford, and John Quarles. *Groundwater Contamination in the United States,* 2d ed. Philadelphia: University of Pennsylvania Press, 1987.

Piller, Charles. *The Fail-Safe Society: Community Defiance and the End of American Technological Optimism.* New York: Basic Books, 1991.

Randolph, Theron G., M.D. *Human Ecology and Susceptibility to the Chemical Environment.* Springfield, Illinois: Charles C. Thomas, Publishers, 1962.

————, and Ralph W. Moss, Ph.D. *An Alternative Approach to Allergies.* New York: Lippincott and Crowell, 1980.

Rathje, William, and Cullen Murphy. *Rubbish: The Archaeology of Garbage.* New York: HarperCollins, 1992.

Schweitzer, Glenn E. *Borrowed Earth, Borrowed Time: Healing America's Chemical Wounds.* New York: Plenum Press, 1991.

Scott, Anne Firor. *Natural Allies: Women's Associations in American History.* Urbana: University of Illinois Press, 1991.

Shulman, Seth. *The Threat at Home: Confronting the Toxic Legacy of the U.S. Military.* Boston: Beacon Press, 1992.

Sontag, Susan. *Illness As Metaphor.* New York: Vintage Books, 1979.

Spitz, Peter H. *Petrochemicals: The Rise of an Industry.* New York: John Wiley & Sons, Inc., 1988.

Stephenson, Neal. *Zodiac: The Eco-Thriller.* New York: The Atlantic Monthly Press, 1988.

Straub, Peter. *Floating Dragon.* New York: The Berkeley Publishing Group, 1982.

Trost, Cathy. *Elements of Risk: The Chemical Industry and Its Threat to America.* New York: Times Books, 1984.

Udall, Stuart L. *The Quiet Crisis.* New York: Avon Books, 1970.

Weir, David. *The Bhopal Syndrome.* San Francisco: Sierra Club Books, 1988.

Whelan, Elizabeth. *Toxic Terror: The Truth About the Cancer Scare.* Ottawa, Illinois: Jameson Books, Ottawa, 1985.

Wohl, Anthony S. *Endangered Lives: Public Health in Victorian Britain.* Cambridge, Massachusetts: Harvard University Press, 1983.

Zeff, Robbin Lee, Marsha Love, and Karen Stults, editors. *Empowering Ourselves: Women and Toxics Organizing.* Arlington, Virginia: Citizens' Clearinghouse for Hazardous Wastes, Inc., 1989.

Reports and Studies

Baker, Dean, M.D., M.P.H., and Sander Greenland, M.S., Dr.P.H. *Stringfellow Health Effects Study: An Epidemiologic Health Survey of Residents of Glen Avon and Rubidoux, California.* Los Angeles: School of Public Health, University of California, 1986.

————, James Mendlein, Ph.D., and Patricia Harmon, M.P.H. "A Health Study of Two Communities Near the Stringfellow Waste Disposal Site," *Archives of Environmental Health.* Vol. 34, No. 5, September/October 1988.

California Department of Health Services. *Pregnancy Outcomes in Santa Clara County 1980–1982: Reports of Two Epidemiological Studies.* Berkeley: Epidemiological Studies Section, 1985.

————, Environmental Epidemiology and Toxicology Program. *McFarland Community Health Newsletter.* Vol. 1, No. 2, October 1, 1988.

————. *Questions and Answers about the McFarland Cancer Cluster.* October 1990.

————, Environmental Epidemiology and Toxicology Program. "Summary of Environmental Data: McFarland Childhood Cancer Cluster Investigation." Phase III Report, October 1991.

————. *History of the McFarland Health Studies.* October 1991.

Center for Third World Organizing. *Toxics and Minority Communities.* The Alternative Policy Institute of the Center for Third World Organizing, July 1986.

Chemical Waste Management, Inc. *Solid and Hazardous Waste: A Basic Study Guide.* Montgomery, Alabama, 1991.

Chess, Caron, Billie Jo Hance, and Peter M. Sandman. *Improving Dialogue with Communities: A Short Guide for Government Risk Communication.* Trenton: New Jersey Department of Environmental Protection, 1988.

Citizen's Clearinghouse for Hazardous Wastes. *Community Health Surveys.* Falls Church, Virginia.

Cohen, David B., Ph.D, and Gerald W. Bowes, Ph.D., California State Water Resources Control Board. *Water Quality and Pesticides: A California Risk Assessment Program.* Vol. 1. Sacramento: December 1984.

Committee for Racial Justice, United Church of Christ. *Toxic Wastes and Race in the United States: A National Report on the Racial and Socio-Economic Characteristics of Communities with Hazardous Waste Sites.* Cleveland: 1987.

Coye, Molly, M.D., M.P.H., and Lynn R. Goldman, M.D., M.P.H. *The Four County Study of Childhood Cancer Incidence Interim Report #2: DRAFT.* Emeryville, California: California Department of Health Services, Environmental Epidemiology and Toxicology Program, October 24, 1991.

Elliot, Donald Jr. "Environmental Equity in the 1990s: Pollution, Poverty and Political Empowerment." *The Kansas Journal of Law and Public Policy,* Vol. 1, No. 1, 1991.

Environmental Protection Agency. *Disposal of Hazardous Wastes.* Report to Congress, Publication SW-115, 1974.

————. *Environmental Protection Agency's Management of the Superfund Program— An Overview.* December 1988, A Report to The Committee on Appropriations, U.S. House of Representatives.

————. *Reducing Risk for All Communities.* Report to the Administrator, Equity Workshop, February 1992.

Environmental Research Foundation. *Press Reports About Violations of Law by Waste Management, Inc..* Washington, D.C., July 1988.

————. *Landfill Package.* Washington, D.C., July 1988.

Kenna, H. Gordan. *Hazardous Waste: What It Is and Where It Goes.* Southern Region Community Relations Manager, Chemical Wastes Management, Inc.

Lagakos, S.W., B.J. Wessen, and M. Zelen. *The Woburn Study: An Analysis of Reproductive and Childhood Disorders and Their Relation to Environmental Contamination.* Harvard University School of Public Health, February 7, 1984.

Lewis, Sanford, and Marco Kaltofen. *From Poison to Prevention: A White Paper on Replacing Hazardous Waste Facility Sites with Toxics Reduction.* Boston: The National Toxics Campaign, 1989.

Lockheed Engineering and Sciences Company. "X-Ray Fluorescence Site Screening Survey, Soil Sampling, and Chemical Analyses, McFarland, California," Las Vegas, Nevada, report to Office of Research and Development, U.S. Environmental Protection Agency, December 1990.

National Solid Waste Management Association. *Transporting Hazardous Wastes: A Situation Under Control.* Washington, D.C., 1989.

Office of Technology Assessment. *Are We Cleaning Up: 10 Superfund Case Studies.* Washington, D.C.: Special Report, OTA-ITE-362, U.S. Government Printing Office, June 1988.

Paigen, Beverly, and Lynn Goldman. "Prevalence of Health Problems in Children Living Near Love Canal." *Hazardous Waste and Hazardous Materials.* Vol. 2, No. 1, 1985.

Phillips, Amanda, M.P.H., and Ellen Silbergeld, Ph.D. "Health Effects Studies of Exposure from Hazardous Waste Sites: Where Are We Today?" *American Journal of Industrial Medicine,* 1985.

Russell, Dick, Sanford Lewis, and Brian Keating. "Inconclusive By Design: Waste, Fraud, and Abuse in Federal Environmental Health Research." Boston: Environmental Health Network and National Toxics Campaign, 1992.

Sandman, Peter M. *Explaining Environmental Risk: Some Notes on Environmental Risk Communication.* Washington, D.C.: Environmental Protection Agency, Office of Toxic Substances, 1986.

State of California Department of Health Services. "Contaminated Drinking Water in California An Emerging Public Health Concern." *Tox-Epi Review,* December 1985.

State of California Health and Welfare Agency and Department of Health Services. *Final Report on a Monitoring Program for Organic Chemical Contamination of Large Public Water Systems in California.* 1986.

State of California Water Resources Control Board. *Groundwater Contamination by Pesticides: A California Assessment.* Publication No. 83-4SP, June 16, 1983.

Upton, Arthur C., et al. "Public Health Aspects of Toxic Chemical Disposal Sites," *Annual Review of Public Health*. Vol. 10, 1989.

Warner, Stephanie, C., Ph.D., and Timothy E. Eldrich, Ph.D. "The Status of Cancer Cluster Investigations Undertaken by State Health Departments," *American Journal of Health*. Vol. 78, No. 3, March 1988.

INDEX

363.738
Setter- Setterberg, Fred
berg
Toxic nation